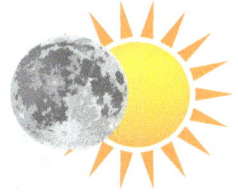

VERMONT
TOTAL ECLIPSE GUIDE

Official Commemorative
2024 Keepsake Guidebook

2024 Total Eclipse State Guide Series

Aaron Linsdau

SASTRUGI PRESS

JACKSON HOLE

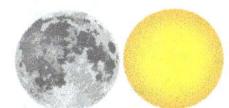

Sastrugi Press / Published by arrangement with the author

Vermont Total Eclipse Guide: Official Commemorative 2024 Keepsake Guidebook

The author has made every effort to accurately describe the locations contained in this work. Travel to some locations in this book is hazardous. The publisher has no control over and does not assume any responsibility for author or third-party websites or their content describing these locations, how to travel there, nor how to do it safely. Refer to local regulations and laws.

Any person exploring these locations is personally responsible for checking local conditions prior to departure. You are responsible for your own actions and decisions. The information contained in this work is based solely on the author's research at the time of publication and may not be accurate in the future. Neither the publisher nor the author assumes any liability for anyone climbing, exploring, visiting, or traveling to the locations described in this work. Climbing is dangerous by its nature. Any person engaging in mountain climbing is responsible for learning the proper techniques. The reader assumes all risks and accepts full responsibility for injuries, including death.

Sastrugi Press
PO Box 1297, Jackson, WY 83001, United States
www.sastrugipress.com
Quantity sales: Special discounts are available on quantity purchases by corporations, associations, and others. For details, contact the publisher at the address above.

Library of Congress Catalog-in-Publication Data
Library of Congress Control Number: 2019949784
Linsdau, Aaron
Vermont Total Eclipse Guide / Aaron Linsdau-1st United States edition
p. cm.
1. Nature 2. Astronomy 3. Travel 4. Photography
Summary: Learn everything you need to know about viewing, experiencing, and photographing the total eclipse in Vermont on April 8, 2024.

ISBN-13: 978-1-944986-33-9 (paperback)

508.4—dc23

Printed in the United States of America when purchased in the United States

All photography, maps and artwork by the author, except as noted.
00063
10 9 8 7 6 5 4 3 2 1

Contents

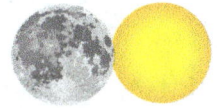

Introduction

Thank you for purchasing this book. It has everything you need to know about the total eclipse in Vermont on April 8, 2024.

A total eclipse passing through the United States is a rare event. The last US total eclipse was in 2017. It traveled from Oregon to South Carolina. The last American total eclipse prior to that was in 1979!

The next total eclipse over the US will not be until April 8, 2024. It will pass over Texas, the Midwest, and on to Maine. After that, the next coast-to-coast total eclipse will be in 2045.

It's imperative to make travel plans early. You will be amazed at the number of people swarming to the total eclipse path. Some might say watching a partial versus a total eclipse is a similar experience. It's not.

This book is written for Vermont visitors and anyone else viewing the eclipse. You will find general planning, viewing, and photography information inside. Should you travel to the eclipse path in Vermont in April, be prepared for an epic trip. The estimates based on the 2017 eclipse suggest that millions will converge on Vermont.

Some hotels in the communities and cities along the path of totality in Vermont have already been contacted by people to make reservations. Finding lodging along the eclipse path may be a major challenge.

Resources will be stretched far beyond the normal limits. Think gas lines from the late 1970s. It may be likely that traffic along highways will come to a complete standstill during this event. Be prepared with backup supplies.

Many smaller Vermont towns are far from any major city. Vermont country roads can be slow. Please obey posted speed limits for the safety of everyone. Be cautious about believing a map application's estimate of travel time in Vermont.

People in communities along the path of the total eclipse may rent out properties for this event. With this major celestial spectacle in the spring of 2024, be assured that Vermont "hasn't seen anything yet."

Is this to say to avoid Vermont or other areas during the eclipse? Not at all! This guidebook provides ideas for interesting, alternative, and memorable locations to see the eclipse. It will be too late to rush to a better spot once the eclipse begins. Law enforcement will be out to help drivers reconsider speeding.

Please be patient and careful. There will be a large rush of people from all over the world, converging on Vermont to enjoy the total eclipse. Be mindful of other drivers on eclipse weekend, as they may not be familiar with Vermont roads.

You should feel compelled to play hooky on April 8. Ask for the day off. Take your kids out of school. They'll be adults before the next chance to see a total eclipse over America. Create family memories that will last a lifetime. Sastrugi Press does not normally advocate skipping school or work. Make an exception because this is too big an event to miss.

Wherever you plan to be along the total eclipse path, leave early and remember your eclipse glasses. People from all around the planet will converge on Vermont. Be good to your fellow humans and be safe. We all want to enjoy this spectacular show.

Visit www.sastrugipress.com/eclipse for the latest updates for this state eclipse book series.

AUTHOR INFORMATION

Polar explorer and motivational speaker Aaron Linsdau's first book, *Antarctic Tears*, is an emotional journey into the heart of Antarctica. He ate two sticks of butter every day to survive. Aaron coughed up blood early in the expedition and struggled with equipment failures. Despite the endless difficulties, he set a world record for surviving the longest solo expedition to the South Pole.

Aaron teaches how to build resilience to overcome adversity by managing attitude. He shares his techniques for overcoming adrenaline burnout and constant overload. He inspires audiences to face their challenges with a new perspective. As a motivational speaker, Aaron talks about courage, resilience, attitude, safety, and risk. He hopes that you will be inspired and have an enjoyable time watching the total eclipse in Vermont. Visit www.aaronlinsdau.com to learn more.

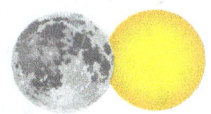

VERMONT

All About Vermont

OVERVIEW OF VERMONT

Vermont, the Green Mountain State, has long been a favorite place for those seeking an escape from stress and relaxation in a quiet rural setting. Tucked between upstate New York and New Hampshire, Vermont's many cities and towns define the idea of the quaint New England village. From its covered bridges and small dairy farms to its infamously green mountains spanning the length of the state, Vermont has plenty to offer people of all ages.

When many people think of Vermont, dairy farms are one of the first things to come to mind. And for a good reason, the state's many dairy farms offer visitors a variety of fresh, locally-sourced dairy goods such as milk, cheese, and of course, ice cream. Travelers to Vermont can visit and tour Ben and Jerry's ice-cream factory. Visitors to

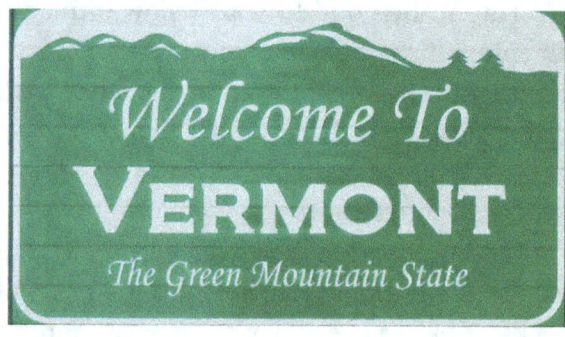

the Vermont-based ice cream makers can tour facilities, sample new flavors, and visit the flavor graveyard, where they can sample flavors that have been long out of production.

Just as iconic as Vermont's dairy farms is its world-famous maple syrup. Vermont's seasonal changes with warm days and cold nights have made it the nation's leader in maple syrup production. And the state offers no shortage of places where the freshly made delicious syrup can be poured on all your favorite breakfast dishes. Family-owned farms and stores like Sugarbush Farms in Woodstock, Vermont, offer fresh maple syrup along with other favorites like local cheese and fruit preserves.

American history lovers will also be at home in Vermont. The

Coolidge Homestead in Plymouth, Vermont, is the boyhood home of America's thirtieth president, Calvin Coolidge. Also known as "Silent Cal," you can visit the exact spot the then Vice President Coolidge took the presidential oath after the sudden death of President Harding. Coolidge's father and owner of the homestead was a judge and swore in his son right on the farm. The Coolidge Homestead offers an interesting look at the humble beginnings behind one of America's presidents.

Outdoor lovers will be spoiled with choices when visiting the Green Mountain State. Vermont's mountains boast plenty of scenic hiking trails for people of all skill levels. Cities like Stowe offer plenty of amazing skiing and hiking trails with some of the most beautiful views of the state's mountains. Many consider Stowe to be the "European Alps of New England." In addition to great hiking, Stowe also offers tourists two of Vermont's best waterfalls. Bingham Falls and Moss Glen Falls are awe-inspiring favorites for people of all ages. Located in Vermont's far north, Stowe is an ideal location for viewing the 2024 eclipse.

And Stowe isn't the only town worth visiting in Vermont. Burlington is the most populated city in the state and is a guaranteed good time. Located on the stunning Lake Champlain, Burlington offers plenty of activities for all tastes. Whether you prefer fishing, hiking, or strolling through downtown, you will not be disappointed. The downtown area offers farmer's markets, restaurants, and novelty stores. The newly renovated Fleming Art Museum is also a must-see for those who will appreciate its world-class collection of fine art and cultural artifacts, including a mummy still in its original coffin.

If you do go to Burlington, taking some time away from the city to visit Lake Champlain is certainly a good decision. The lake, which has no shortage of cabins for rent on its shores, is a favorite of locals and tourists alike. Visitors can rent a boat, canoe, or kayak and explore the vast lake, which has no shortage of fish and stunning views. Just relaxing on the sandy beaches is sure to wash the stress away.

And all this only scratches the surface of what Vermont has to offer. Visitors will also take breathtaking rides on Vermont's many scenic roadways. The small amount of light pollution in Vermont's rural skies will leave you to stargaze the nights away. Vermont's abundance of craft breweries offers visitors plentiful craft beer options, and family-owned restaurants will leave your belly full of Vermont's rustic offerings. In addition to Vermont's great location for viewing the eclipse, there is no shortage of other reasons that will make a trip to Vermont's rural landscape an unforgettable experience for you and your family.

HOTELS AND MOTELS DURING THE ECLIPSE

Once excitement of the total eclipse over Vermont spreads, rooms will become scarce. Many hotels in towns along the path of totality in western states sold out for a year or more during the 2017 total eclipse. Vermont is not alone in this challenge. Hotels all along the path of totality will sell out in anticipation of the 2024 total eclipse.

What does this mean for eclipse visitors? Lodging and room rentals in eclipse towns will be at a massive premium. Does that mean all hope is lost to find a place to stay? Not at all. But you will have to be creative. There will be few, if any, hotel rooms available in these eclipse cities by early 2024. Accommodations in the cities and towns along the path of the eclipse will be difficult to come by.

In summer 2017, the author searched on Hotels.com for rooms along the 2017 total eclipse path on the weekend of August 21 and found many major cities sold out. Once word of the 2024 eclipse spreads, room rates will increase and availability will drop.

Search for rooms farther away from the eclipse path. If you are willing to stay in cities outside the eclipse path, you will have better success at finding rooms. As the eclipse approaches, people will book rooms farther from the totality path. By early spring, rooms in cities near the total eclipse path may be unavailable. The effect of this event will be felt across Vermont and the rest of the United States.

Think regionally when looking for rooms. Be prepared to search far and wide during this major event. If a five-hour drive is manageable, your lodging options greatly expand, but it also increases your travel risk.

INTERNET RENTALS

To find rooms to stay in towns along the eclipse path, try a web service such as Airbnb.com. Note that some people rent out rooms or homes illegally, against zoning regulations. Cities will feel the crunch of inquiries early due to others who experienced the 2017 eclipse.

If cities fully enforce zoning laws, authorities may prevent your weekend home rental. Online home rentals during the eclipse will be a target for rental scams. People from out of the area steal photos and descriptions, then post the home for rent. You send your check or wire money to a "rental agent" then show up to find you have been scammed. If the deal sounds strange or too good to be true, run away.

CAMPING

If you can book a campsite, do it as soon as you can. Do not wait. All areas in the national forests are first-come, first-served. Forest roads may be packed. Expect all areas to be swarming with people. Show up early to stake out your spot. Consider staying farther away and driving early on April 8.

Please respect private land too. Vermont folks don't take kindly to people overrunning their property without permission. In a big state with millions of residents, people are very protective, but they're friendly, too. You never know what you might be able to arrange with a smile and a bit of money.

This all said, there are plenty of camping opportunities throughout Vermont. You don't have to sleep exactly on the eclipse path. If you're ready to rough it, there are national forest camping options.

Government agencies will meet years in advance to talk about how to manage the influx of people. Every possible government agency will be working full time to enforce the various rules and regulations.

NATIONAL PARKS AND MONUMENTS

Finding a camping site at any state park, national park, or national monument along the eclipse path in Vermont will be challenging. To watch the eclipse from any location, you do not have to sleep in it. You just need to drive there in the morning.

Law enforcement will be present on the eclipse weekend. Hundreds of thousands of people are expected in the region. Parking may overflow. It will make parking lots and lines on Black Friday at the mall look uncrowded. For an event of this magnitude, find your location as early as possible.

The first sentence of the national parks mission statement is:

"The National Park Service preserves unimpaired the natural and cultural resources and values of the national park system for the enjoyment, education, and inspiration of this and future generations."

Roadside camping (sleeping in your car) is not allowed in national monuments or parks. Park facilities are only designed to handle so many people per day. Water, trash collection, and toilets can only withstand so much. If you notice trash on the ground, take a moment to throw it away. Protect your national park and help out. Rangers are diligent and hardworking but they can only do so much to manage the expected crowds.

National and State Forests

There are national and state forest options in Vermont. They all have camping opportunities. The forest service manages undeveloped and primitive campsites. Be sure to check for any fire restrictions. Check with individual agencies for last-minute information and regulations. The forest service requires proper food storage. Plan to purchase food and water before choosing your campsite. Below is a partial list of forests and parks in Vermont:

Green Mountain National Forest: https://www.fs.usda.gov/gmfl
Jay State Forest
Long Trail State Forest
Mount Mansfield State Forest

Backcountry service roads abound in Vermont. Maps for forests are available at local visitor centers and bookstores. This book's website has digital copies of some forest maps.

Printed national forest maps are large and detailed. They have illustrated road paths, connections, and other vital travel information not available on digital device maps. Viewing digital maps on your smartphone or mobile pad is difficult. If you plan to camp in the forest, a real paper map is a wise investment.

Camping in federal wilderness areas is also allowed. Those areas afford the ultimate backcountry experience. However, be aware that no vehicle travel is allowed in the specially designated areas. This ban includes: vehicles, bikes, hang gliders, and drones. You can travel only on foot or with pack animals.

SLEEP IN YOUR CAR

Countless RVs, campers, trucks, cars, and motorcycles will flood Vermont. Sleeping in your car with friends is tolerable. Doing so with unadventurous spouses or children is another matter.

Do not be caught along the path of the total eclipse without some sort of plan, especially in the bigger cities of Vermont. The whole path of totality will fill with people on April 8.

USEFUL LOCAL WEBCAMS

Local webcams are handy to make last-minute travel decisions. The webcams are sensitive enough to show headlights at night. Use them to determine if there are issues before traveling out. Eclipse traffic will add to the morning commuter traffic.

The smartphone application Wunderground is useful to check on webcams in one place. Selected the webcams are listed in the app. Whether you use this app or another, an Internet search will reveal many useful webcams for your search.

Weather

It's all about the weather during the eclipse. Nothing else will matter if the sky is cloudy. You can be nearly anywhere along the path in Vermont and catch a view of the event when traffic comes to a standstill. But if there's a cloud cover forecast, seriously reconsider your viewing location.

Travel early wherever you plan to go. Attempting to change locations an hour before the eclipse due to weather will likely cause you to miss the event. Vermont country roads can be narrow and slow. The number of vehicles will cause unexpected backups.

Modern Forecasts

Use a smartphone application to check the up-to-date weather. Wunderground is a good application and has relatively reliable forecasts for the region. The hourly forecast for the same day has been rather accurate for the last two years. The below discussion refers to features found in the Wunderground app. However, any application with detailed weather views will improve your eclipse forecasting skills.

Cloud Cover Forecast

The most useful forecast view is the visible and infrared cloud-coverage map. Avoid downloading this app the night before and trying to learn how to read it. Practice reading them at home. It's imperative to understand how to interpret the maps early.

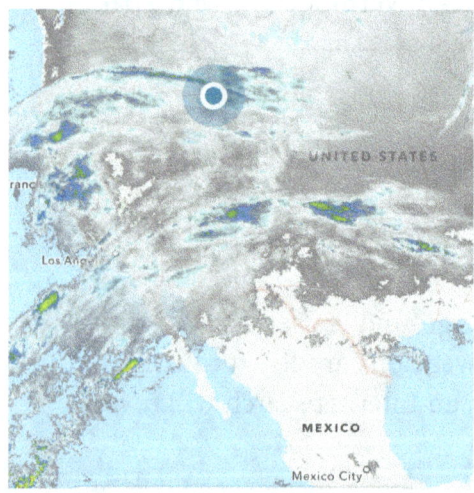

Infrared cloud map showing the worst case eclipse cloud cover.
Courtesy of National Weather Service.

All cloud cover, night or day, will appear on an infrared map. Warm, low-altitude clouds are shown in white and gray. High-altitude cold clouds are displayed in shades of green, yellow, red, and purple. Anything other than a clear map spells eclipse-viewing problems.

To improve your weather guess, use the animated viewer of the cloud cover. It will give you a sense of cloud motion. You can discern whether clouds or rain are moving toward, away from, or circulating around your location.

Normal Vermont Weather Pattern

Due to the direction of the jet stream, most weather travels across the Pacific Ocean, through the western states, over the Great Lakes, and then into Vermont. On occasion, weather can approach from Canada or the Atlantic. Due to the nature of the storms from the

Arctic, weather in Vermont can be unpredictable in the spring.

The common weather pattern in April is slightly warm in the afternoon and cool to cold in the evenings. Passing cold fronts in spring can bring regular cloud cover and rain showers.

Historically, Vermont tends to have moderate cloud cover during April. Prepare to make adjustments. If anything other than clear skies are predicted, drive to other parts of Vermont, New York, or New Hampshire.

Be aware of severe weather in Vermont. Although spring is well underway by the time of the total eclipse, there have been many substantial storms in April. Pay attention to the weather forecast. If dangerous weather is predicted, your main concern should be safety rather than chasing an eclipse.

Consider that slow-moving clouds can obscure the sun for far longer than the local duration of the totality. The time of totality is so short that you do not want to risk it. Missing it due to a single cloud will be a major disappointment.

Local Eclipse Weather Forecasts

Local town and city newspapers, radio, and television stations around Vermont will have a weekend edition with articles discussing the eclipse weather. However, conditions change unpredictably in Vermont. A three-day forecast for April may be incorrect.

Finding the Right Location to View Eclipse Effects

One of the peculiarities of total eclipses is that the entire show is not only in the sky. There are other unusual effects seen during the total eclipse that are worth looking for.

The first effect to watch for is the crescent moon shapes created from leaf shadows on the ground. They're best viewed on a sidewalk or asphalt. They can only be observed during the

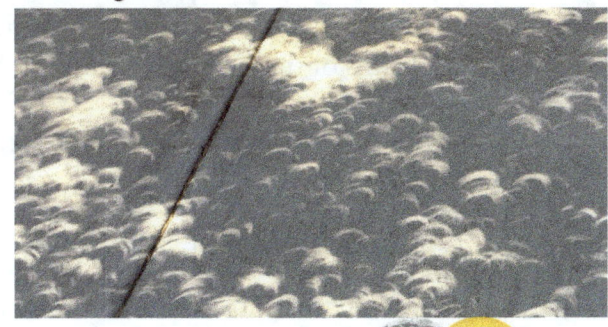

VERMONT

partial eclipse. The other effect that is worth watching for is the shadow bands or "snakes" as they're commonly called.

Shadow banding is seen right before and after the totality takes place. They're best observed on smooth, plain-colored surfaces. If you plan to be in the forest for the eclipse, you may struggle to see the bands but will likely see crescent shadows all around on the ground.

One of the supreme challenges with all of these effects is choosing what to watch. You can see the crescent shadows in the hour before and after the totality but shadow banding happens before or after totality. It is more difficult to look away from the eclipse than you think.

Road Closures and Traffic

Highways connecting various Vermont towns in the total eclipse path will be heavily impacted on the weekend before and day of the eclipse. As was found in the 2017 total eclipse, there is no way to predict which areas will be impacted.

Planning ahead is essential to give you the best opportunity to enjoy the eclipse without the nightmare of being stuck in traffic for hours on end. The traffic in states during the 2017 eclipse was stunning, so it may be substantial in Vermont in 2024.

Update yourself with the latest road report information from the Vermont road condition website:

https://newengland511.org/

It's imperative to plan early and have one if not more backup plans in case of difficult travel conditions. April weather is unpredictable and variable.

If you believe it's necessary to leave a town to watch the eclipse, do so the night before or extremely early in the morning. RVs are common, and trains of them crawl through popular areas.

Communication Information

Cellular Phones

Cellular "cell" phone service in some remote Vermont locations may be problematic. Most of the time there is good coverage along the main highways and interstates. However, even along major thoroughfares, there can be little or no coverage.

It's possible to find zones where text messages will send when phone calls are impossible. If you cannot make a phone call, the chance of having data coverage for web surfing or e-mail is low.

Please look up any information or communicate what you need before departing from the main roads around Vermont. Bureau of Land Management (BLM) areas sometimes have coverage. Planned to be self-contained. Plan for your cell phone not to connect.

You may find yourself out of cell service. With a large number of cell users in a concentrated area, coverage and data speed may collapse as well. Search on the phrase "cell phone coverage breathing".

Wilderness and Forest Safety

All Vermont forest and wilderness areas are full of wild animals. Although beautiful, wild animals can be dangerous. They can easily injure or kill people, as they are far more powerful than humans. Do not try to feed any wild animals, including squirrels, foxes, and chipmunks, as they can carry diseases. These suggestions apply to all public lands.

TICKS

Ticks exist all across the United States, but not all species transmit disease. Ticks cannot fly or jump, but they climb grasses in shrubs in order to attach to people or animals that pass by. Ticks feed on the blood of their host. In doing so, they can transmit potentially life-threatening diseases such as Lyme disease.

SPIDERS

Although the chance of encountering a venomous spider is low, it is not uncommon to encounter them. The two dangerous spiders in the state are the black widow and brown recluse. Should you encounter either species, simply leave it alone. If you are bitten, seek immediate medical attention, as their toxin can be life-threatening.

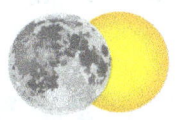

VERMONT

Venomous Snakes

There is one known species of venomous snake in Vermont, the Timber Rattlesnake. Although these reptiles are not generally aggressive and rarely seen, they can strike when provoked or threatened. Of the approximately 8,000 people annually bitten by venomous snakes in the United States, ten to fifteen people die according to the U.S. Food and Drug Administration.

The best way to avoid rattlesnake encounters is to be mindful of your environment. Do not place your hands or feet in locations where you cannot clearly see the surroundings. Avoid heavy brush or tall weeds where snakes hide during the day. Step on a log or rock rather than over it, as a hidden snake might be on the other side. Rattlesnakes may not make any noise before striking.

Avoid handling all snakes. Should you be bitten, stay calm and call 911 or emergency dispatch as soon as possible. Transport the victim to the nearest medical facility immediately. Rapid professional treatment is the best way to manage rattlesnake bites. Refer to US Forest Service and professional medical texts for more information on managing rattlesnakes injuries.

Bears

The forests of Vermont are potentially home to black bears. Though they are listed as not often seen, they have been sighted in the state. Although they often appear docile, they can become aggressive if threatened. In the unlikely event of an attack, fight back against the bear. Use whatever you have at your disposal to defend yourself. Report all negative or aggressive bears to the local authorities.

If a bear hears you, it will usually vacate the area. Bear charges are often caused by unexpected and surprise encounters. Noise is the best defense to avoid surprising bears. Regularly clap, make noise, and talk loudly. The Vermont Fish& Wildlife Department is a good starting point to learn more about black bears at https://vtfishandwildlife. com/learn-more/vermont-critters/mammals/black-bear.

It is recommended to stay one hundred yards (300 feet) away from all bears. They are exciting to see but need their space. Refer to current forest or park regulations for more safety information.

Mountain Lions

Though declared to be extinct in the state, there have been reports of mountain lions by various news agencies. If you encounter a mountain lion, do not run. Keep calm, back away slowly, and maintain eye contact. Do all you can to appear larger. Stand upright, raise your arms, or hoist your jacket. Never bend over or crouch down. If attacked, fight back.

Eclipse Day Safety

1. Hydrate

Spring temperatures are usually mild to warm. The excitement of the event can distract you from managing hydration. Drink plenty of water. Consume more than you would at home.

2. Eye Safety time

Use certified eclipse safety glasses at all times when viewing the partial eclipse. Only remove the glasses when the totality happens. Give your eyes time to rest. They can dry out and become irritated. Bring FDA approved eye drops to keep your eyes moist.

3. Sun exposure

Facing at the sun for three hours can result in sunburns. Wear sunglasses and liberally apply sunscreen to avoid sunburns.

4. Eat well

Keep your energy up. Appetite loss is common when traveling. Maintain your normal eating schedule.

5. Prepare for temperature changes

Temperatures will drop rapidly during the eclipse and also once the sun sets. Bring appropriate clothing.

6. Talk with your doctor

If the humidity or heat bothers you talk with your doctor before traveling. Seek professional medical attention for serious symptoms.

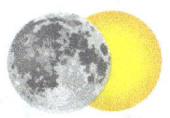

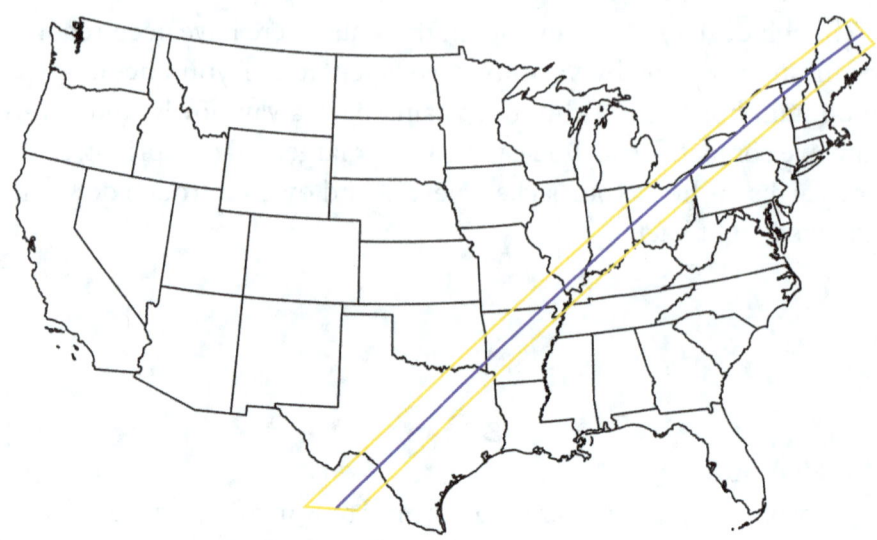

Total eclipse path across the United States (approximate).

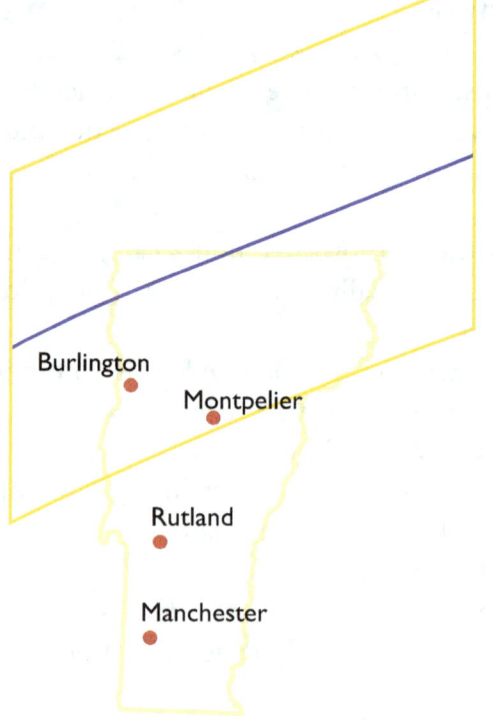

Total eclipse path across Vermont (approximate).

All About Eclipses

How an Eclipse Happens

An eclipse occurs when one celestial body falls in line with another, thus obscuring the sun from view. This occurs much more often than you'd think, considering how many bodies there are in the solar system. For instance, there are over 150 moons in the solar system. On Earth, we have two primary celestial bodies: the sun and the moon. The entire solar system is constantly in motion, with planets orbiting the sun and moons orbiting the planets. These celestial bodies often come into alignment. When these alignments cause the sun to be blocked, it is called an eclipse.

For an eclipse to occur, the sun, Earth, and moon must be in alignment. There are two types of eclipses: solar and lunar. A solar eclipse occurs when the moon obscures the sun. A lunar eclipse occurs when the moon passes through Earth's shadow. Solar eclipses are much more common, as we experience an average of 240 solar eclipses a century compared to an average of 150 lunar eclipses. Despite this, we are more likely to see a lunar eclipse than a solar eclipse. This is due to the visibility of each.

For a solar eclipse to be visible, you have to be in the moon's shadow. The problem with viewing a total eclipse is that the moon casts a small shadow over the world at any given time. You have to be in

ECLIPSES

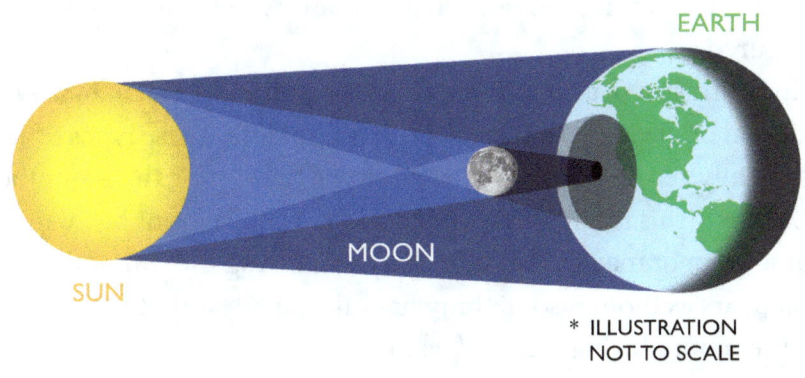

EARTH

MOON

SUN

* ILLUSTRATION
NOT TO SCALE

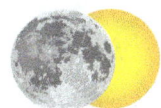

a precise location to view a total eclipse. The issue that arises is that most of these locations are inaccessible to most people. Though many would like to see a total solar eclipse, most aren't about to set sail for the middle of the Pacific Ocean. In fact, a solar eclipse is visible in the same place on the world on average every 375 years. This means that if you miss a solar eclipse above your hometown, you're not going to see another one unless you travel or move.

It's much easier to catch a glimpse of a lunar eclipse, even though they occur at a much lower frequency than their solar counterparts. A lunar eclipse darkens the moon for a few hours. This is different than a new moon when it faces away from the sun. During these eclipses, the moon fades and becomes nearly invisible.

Another result of a lunar eclipse is a blood moon. Earth's atmosphere bends a small amount of sunlight onto the moon turning it orange-red. The blood moon is caused by the dawn or dusk light being refracted onto the moon during an eclipse.

Lunar eclipses are much easier to see. Even when the moon is in the shadow of Earth, it's still visible throughout the world because of how much smaller it is than Earth.

Total vs. Partial Eclipse

What is the difference between a partial and total eclipse? A total eclipse of either the sun or the moon will occur only when the sun, Earth, and the moon are aligned in a perfectly straight line. This ensures that either the sun or the moon is partially or completely obscured.

In contrast, a partial eclipse occurs when the alignment of the three celestial bodies is not in a perfectly straight line. These types of eclipses usually result in only a part of either the sun or the moon being obscured. This is often what led to ancient civilizations believing that some form of magical beast or deity was eating the sun or the moon. It appears as though something has taken a bite out of either the sun or the moon during a partial eclipse.

Total eclipses, rarer than partial eclipses, still occur quite often. It's more difficult for people to be in a position to experience such an event firsthand. Total solar eclipses can only be viewed from a small portion of the world that falls into the darkest part of the moon's shadow. Often this happens in the middle of the ocean.

THE MOON'S SHADOW

The moon's shadow is divided into two parts: the umbra and the penumbra. The former is much smaller than the latter, as the umbra is the innermost and darkest part of the shadow. The umbra is thus the central point of the moon's shadow, meaning that it is extremely small in comparison to the entire shadow. For a total solar eclipse to be visible, you need to be directly beneath the umbra of the moon's shadow. This is because that is the only point at which the moon completely blocks the view of the sun.

In contrast, the penumbra is the region of the moon's shadow in which only a portion of the light cast by the sun is obscured. When

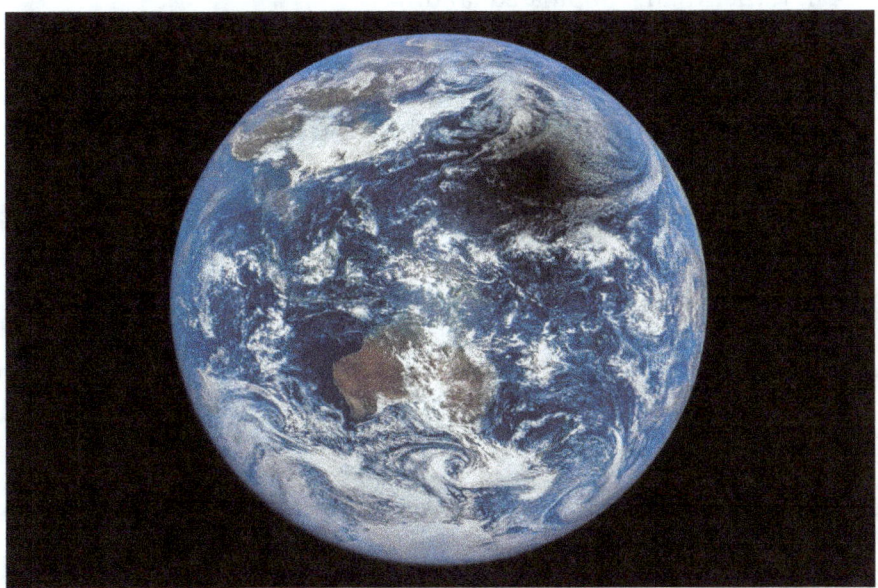

Total eclipse shadow 2016 as seen from 1 million miles on the Deep Space Climate Observatory satellite. Courtesy of NASA.

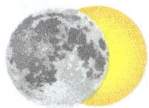

standing in the penumbra, you are viewing the eclipse at an angle. In the penumbra, the moon does not completely block the sun from view. This means that while the event is a total solar eclipse, you'll only see a partial eclipse. The umbra for the April 8 eclipse is over one hundred miles wide. The penumbra will cover much of the United States.

To provide some context, one total solar eclipse we experienced occurred on March 9, 2016, and was visible as a partial eclipse across most of the Pacific Ocean, parts of Asia, and Australia. However, the only place in the world to view this total solar eclipse was in a few parts of Indonesia.

Due to the varied locations and the brief periods for which they're visible, it's difficult to see each and every eclipse that occurs. Many people don't even realize that they have occurred. Consider that the umbra of the moon represents such a small fraction of the entire shadow and the majority of our planet is comprised of water. Thus, the rarity of being able to view a total solar eclipse increases significantly because it's likely that the umbra will fall over some part of the ocean rather than a populated landmass.

Eclipses Throughout History

Ancient peoples believed eclipses were from the wrath of angry gods, portents of doom and misfortune, or wars between celestial beings. Eclipses have played many roles in cultures, creating myths since the dawn of time. Both solar and lunar eclipses affected societies worldwide. Inspiring fear, curiosity, and the creation of legends, eclipses have cast a long shadow in the collective unconscious of humanity throughout history.

Early Myth & Astronomy

Documented observations of solar eclipses have been found as far back in history as ancient Egyptian and Chinese records. Timekeeping was important to ancient Chinese cultures. Astronomical observations were an integral factor in the Chinese calendar. The first observation of a solar eclipse is found in Chinese records from over 4,000 years ago. Evidence suggests that ancient Egyptian observations may predate those archaic writings.

Many ancient societies, including Roman, Greek and Chinese civilizations, were able to infer and foresee solar eclipses from astronomical data. The sudden and unpredictable nature of solar eclipses had a stressful and intimidating effect on many societies that lacked the scientific insight to accurately predict astronomical events. Relying on the sun for their agricultural livelihood, those societies interpreted solar eclipses as world-threatening disasters.

In ancient Vietnam, solar eclipses were explained as a giant frog eating the sun. The peasantry of ancient Greece believed that an eclipse was the sign of a furious godhead, presenting an omen of wrathful retribution in the form of natural disasters. Other cultures were less speculative in their investigations. The Chinese Song Dynasty scientist Shen Kuo proved the spherical nature of the Earth and heavenly bodies through scientific insight gained by the study of eclipses.

The Eclipse in Native American Mythology

Eclipses have played a significant role in the history of the United States. Before Europeans settled in the Americas, solar eclipses were important astronomical events to Native American cultures. In most native cultures, an eclipse was a particularly bad omen. Both the sun and the moon were regarded as sacred. Viewing an eclipse, or even being outside for the duration of the event, was considered highly taboo by the Navajo culture. During an eclipse, men and women would simply avert their eyes from the sky, acting as though it was not happening.

The Choctaw people had a unique story to explain solar eclipses. Considering the event as the mischievous actions of a black squirrel and its attempt to eat the sun, the Choctaw people would do their best to scare away the cosmic squirrel by making as much noise as possible until the end of the event, at which point cognitive bias

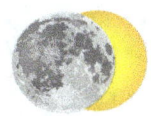

would cause them to believe they'd once again averted disaster on an interplanetary scale.

Contemporary American Solar Phenomena

The investigation of solar phenomena in twentieth-century American history had a similarly profound effect on the people of the United States. A total solar eclipse occurring on the sixteenth of June, 1806, engulfed the entire country. It started near modern-day Arizona. It passed across the Midwest, over Ohio, Pennsylvania, New York, Massachusetts, and Connecticut. The 1806 total eclipse was notable for being one of the first publicly advertised solar events. The public was informed beforehand of the astronomical curiosity through a pamphlet written by Andrew Newell entitled *Darkness at Noon, or the Great Solar Eclipse.*

This pamphlet described local circumstances and went into great detail explaining the true nature of the phenomenon, dispelling myth and superstition, and even giving questionable advice on the best methods of viewing the sun during the event. Replete with a short historical record of eclipses through the ages, the *Darkness at Noon* pamphlet is one of the first examples of an attempt to capitalize on the mysterious nature of solar eclipses.

Another notable American solar eclipse occurred on June 8, 1918. Passing over the United States from Washington to Florida, the eclipse was accurately predicted by the U.S. Naval Observatory and heavily documented in the newspapers of the day. Howard Russell Butler, painter and founder of the American Fine Arts Society, painted the eclipse from the U.S. Naval Observatory, immortalizing the event in *The Oregon Eclipse.*

Four more total solar eclipses occurred over the United States in the years 1923, 1925, 1932, and 1954, with another occurring in 1959. The October 2, 1959, solar eclipse began over Boston, Massachusetts. It was a sunrise event that was unviewable from the ground level. Eminent astronomer Jay Pasachoff attributed this event to sparking his interest in the study of astronomy. Studying under Professor Donald Menzel of Williams College, Pasachoff was able to view the event from an airline hired by his professor.

To this day, many myths surround the eclipse. In India, some local customs require fasting. In eastern Africa, eclipses are seen as a danger to pregnant women and young children. Despite the mystery and legend associated with unique and rare astronomical events, eclipses continue to be awe-inspiring. Even in the modern day, eclipses draw out reverential respect for the inexorable passing of celestial bodies. They are a reminder of the intimate relationship between the denizens of Earth and the universe at large.

Present Day Eclipses

The year 2017 brought the world's most-watched total eclipse in history on August 21, when a total solar eclipse crossed the United States. An annular eclipse, a "ring of fire," will pass over the United States in 2023 from Oregon to Texas. Though impressive, it will not compare to the 2024 total eclipse. There is little in nature that equals

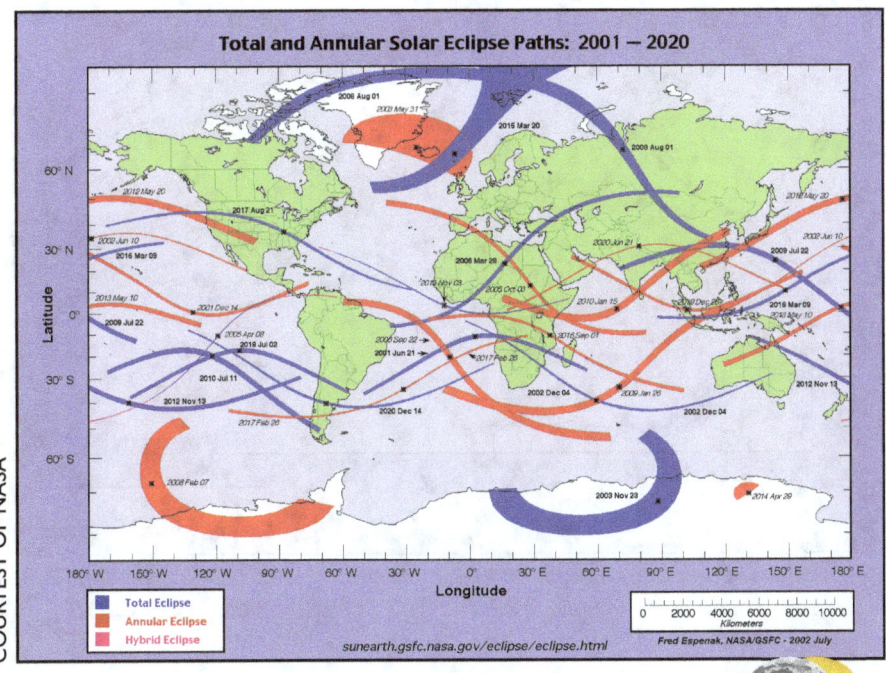

COURTESY OF NASA

ECLIPSES

the spectacle of the sun's corona and seeing stars in the day.

There will be multiple partial, annular, or hybrid eclipses across the world before the 2024 total eclipse. However many are in remote, inaccessible, or potentially dangerous locations on the globe. In 2019 and 2020, Chile and Argentina will experience total eclipses. The next total eclipse after that will occur over Antarctica in 2021. An extremely rare hybrid eclipse will happen in 2023 over the Indian Ocean, Australia, and Indonesia.

The next total solar eclipse viewable from the United States will occur on April 8, 2024. It will be visible in fifteen states: Texas, Oklahoma, Arkansas, Missouri, Tennessee, Kentucky, Illinois, Indiana, Ohio, Pennsylvania, Michigan, New York, Vermont, New Hampshire, and Maine.

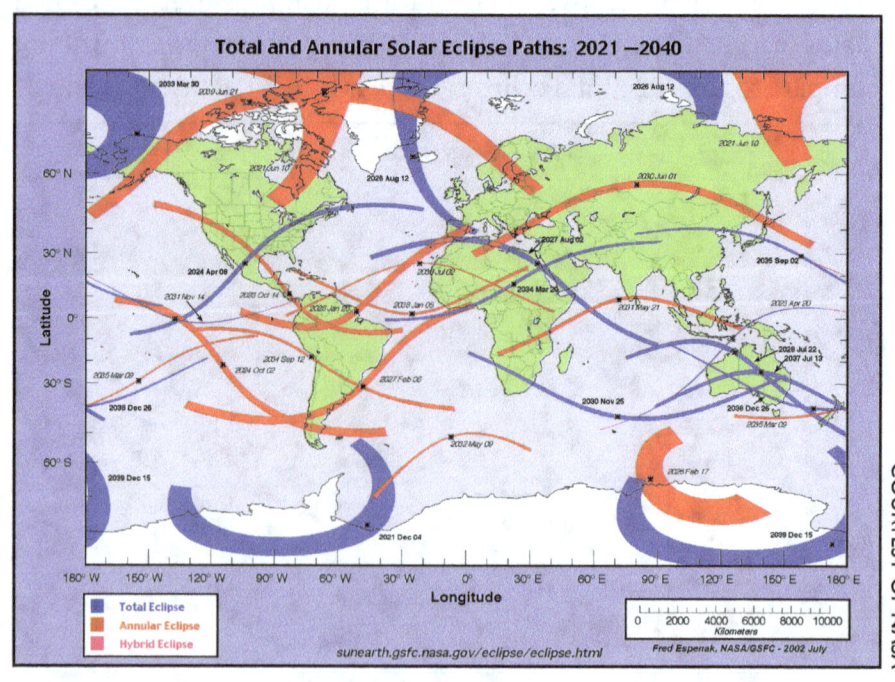

COURTESY OF NASA

Viewing and Photographing the Eclipse

AT-HOME PINHOLE METHOD

Use the pinhole method to view the eclipse safely. It costs little but is the safest technique there is. Take a stiff piece of single-layer cardboard and punch a clean pinhole. Let the sun shine through the pinhole onto another piece of cardboard. That's it!

Never look at the sun through the pinhole. Your back should be toward the sun to protect your eyes. To brighten the image, simply move the back piece of cardboard closer to the pinhole. To see it larger, move the back cardboard farther away. Do not make the pinhole larger. It will only distort the crescent sun.

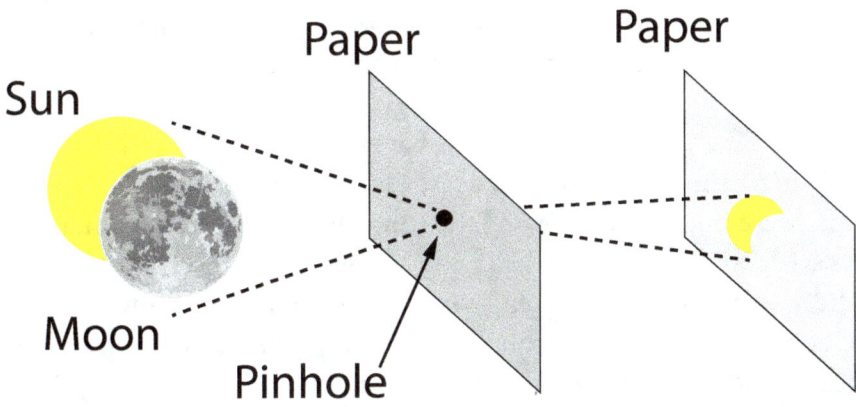

WELDING GOGGLES

Welding goggles that have a rating of fourteen or higher are another useful eclipse viewing tool. The goggles can be used to view the solar eclipse directly. Do not use the goggles to look through binoculars or telescopes, as the goggles could potentially shatter due to intense direct heat. Avoid long periods of gazing with the goggles. Look away every so often. Give your eyes a break.

SOLAR FILTERS FOR TELESCOPES

The ONLY safe way to view solar eclipses using telescopes or binoculars is to use solar filters. The filters are coated with metal

PHOTOGRAPHY

to diminish the full intensity of the sun. Although the filters can be expensive, it is better to purchase a quality filter rather than an inexpensive one that could shatter or melt from the heat.

The filters attach to the front of the telescope for easy viewing. Remember to give your telescope cooling breaks. Rapid heating can damage your equipment with or without filters attached.

WATCH OUT FOR UNSAFE FILTERS

There are several myths surrounding solar filters for eclipse viewing. In order for filters to be safe, they must be specially designed for looking at a solar eclipse. The following are all unsafe for eclipse viewing and can lead to retinal damage: developed colored or chromogenic film, black-and-white negatives such as X-rays, CDs with aluminum, smoked glass, floppy disk covers, black-and-white film with no silver, sunglasses, or polarizing films.

WATCH OUT FOR UNSAFE ECLIPSE GLASSES

During the 2017 total eclipse, several vendors sold eclipse glasses that were not safe for viewing the sun. Although they were marketed as safe and were even marked with the ISO 12312-2 certification, they did not block eye-damaging visible, infrared, and ultraviolet light. Check the American Astronomical Society's website (eclipse.aas.org) for a list of reputable eclipse glasses vendors.

VIEWING WITH BINOCULARS

When viewing the eclipse with binoculars, it is important to use solar filters on both lenses until totality. Only then is it safe to remove the filter. As the sun becomes visible after totality, replace the filters for safe viewing. Protect your pupils. Remember to give your binoculars a cool-down break between viewings. They can overheat rapidly from being pointed directly at the sun even with filters attached.

PLANNING AHEAD

There are many things to keep in mind when viewing a total eclipse. It is important to plan ahead to get the most out of this extraordinary experience.

PHOTOGRAPHY

UNDERSTANDING SUN POSITION

All compass bearings in this book are true north. All compasses point to Earth's magnetic north. The difference between these two measurements is called magnetic declination. The magnetic declination for Vermont is:

13° 56' W ± 0° 23' (for Montpelier in 2024)

Adjust the declination from the azimuth bearing as given in the text, and set your compass to that direction.

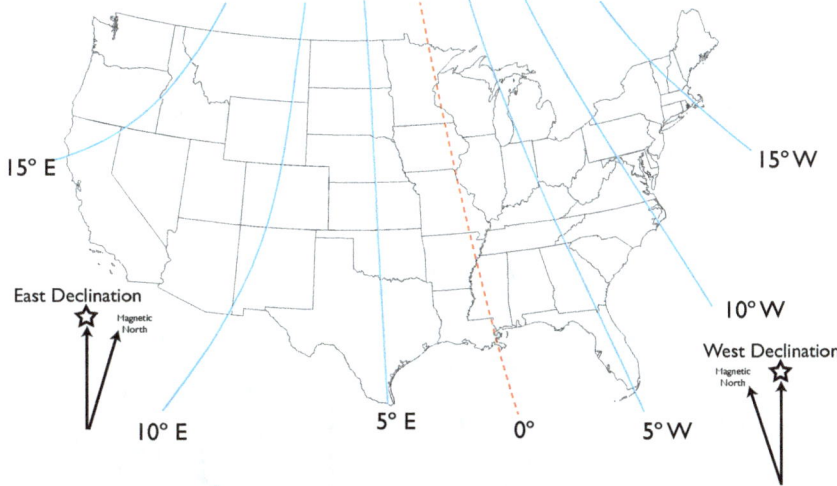

If you purchase a compass with a built-in declination adjustment, you can change the setting once and eliminate the calculations. The Suunto M-3G compass has this correction. A compass with a sighting mirror or wire will help you make a more accurate azimuth sighting.

The Suunto M-3G also has an inclinometer. This allows you to measure the elevation of any object above the horizon. Use this to figure out how high the sun will be above your position.

You can also use a smartphone inclinometer and compass for this purpose. Make sure to calibrate your smartphone's compass before every use, otherwise it might indicate the wrong bearing. Set the smartphone compass for true north to match the book. Understand the compass prior to April 8. There will be little time to guess or search on the Internet. Smartphone and GPS compasses are "sticky." Their compasses don't swing as freely as a magnetic compass does.

The author has used his magnetic compass for azimuth measurements and a smartphone to measure elevation. Combining these two tools will allow you to make the best sightings possible.

Outdoor sporting goods stores in most towns and cities carry compasses. Purchase and practice with a good compass in your hometown well before the event. Take the time to learn how to use it before the day of the eclipse. You do not want to struggle with orienteering basics under pressure.

SUN AZIMUTH

Azimuth is the compass angle along the horizon, with 0° corresponding to north, and increasing in a clockwise direction. 90° is east, 180° is south, and 270° is west.

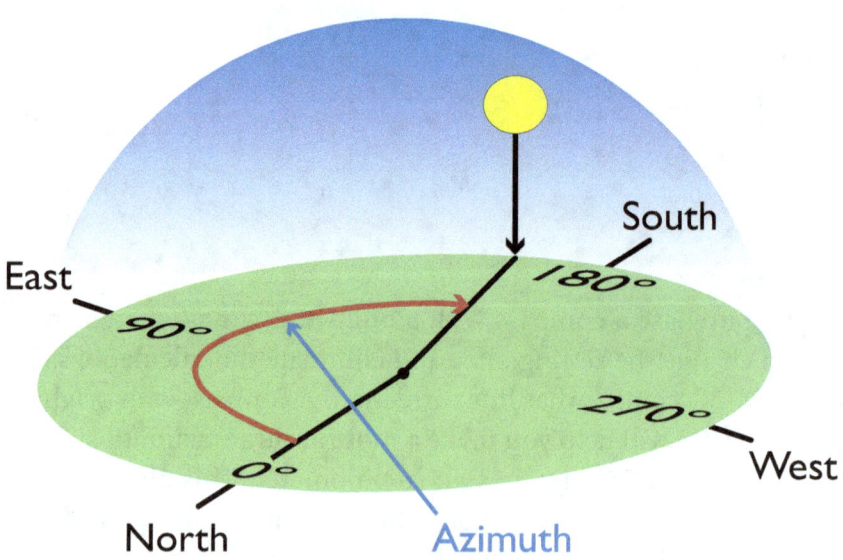

SUN ELEVATION

Altitude is the sun's angle up from the horizon. A 0° altitude means exactly on the horizon and 90° means "straight up."

PHOTOGRAPHY

Using the sun azimuth and elevation data, you can predict the position of the sun at any given time. Positions given in this book coincide with the time of eclipse totality unless otherwise noted.

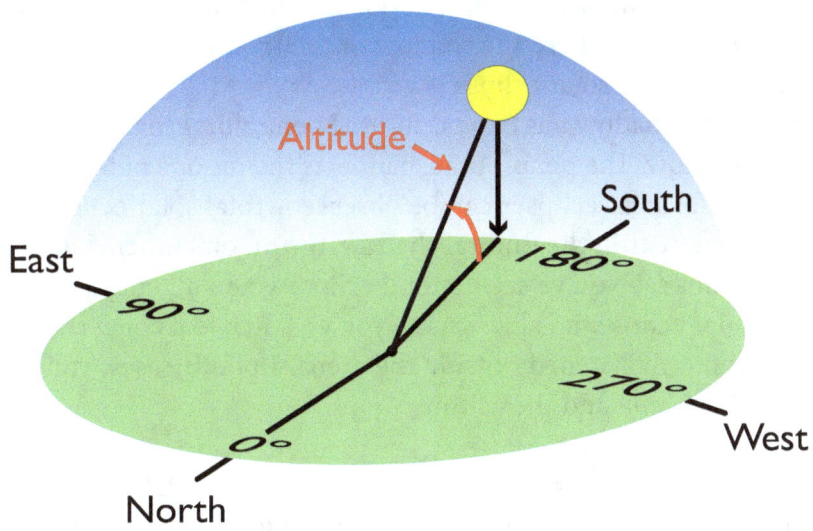

ECLIPSE DATA FOR SELECT VERMONT LOCATIONS

LOCATION	TOTALITY START (EDT)
BURLINGTON	3:26:05PM
EDEN	3:26:53PM
MIDDLEBURY	3:27:10PM
MONTPELIER	3:27:36PM
NEWPORT	3:27:14PM
SAINT JOHNSBURY	3:28:19PM

ECLIPSE PHOTOGRAPHY

Photographing an eclipse is an exciting challenge, as the moon's shadow moves near 1,600MPH. There is an element of danger and the pressure of time. Looking at the unfiltered sun through a camera can permanently damage your vision and your camera. If you are unsure, just enjoy the eclipse with specially designed eclipse glasses. Keep a solar filter on your lens during the eclipse and remove for the duration of totality!

PHOTOGRAPHY

PARTIAL VS. TOTAL SOLAR ECLIPSE

To successfully and safely photograph a partial and total eclipse, it is important to understand the difference between the two. A solar eclipse occurs when the moon is positioned between the sun and Earth. The region where the shadow of the moon falls upon Earth's surface is where a solar eclipse is visible.

The moon's shadow has two parts—the penumbral shadow and the umbral shadow. The penumbral shadow is the moon's outer shadow where partial solar eclipses can be observed. Total solar eclipses can only be seen within the umbral shadow, the moon's inner shadow.

You cannot say you've seen a total eclipse when all you saw was a partial solar eclipse. It is like saying you've watched a concert, but in reality, you only listened outside the arena. In both cases, you have missed the drama and the action.

PHOTOGRAPHING A PARTIAL AND TOTAL SOLAR ECLIPSE

Aside from the region where the outer shadow of the moon is cast, a partial solar eclipse is also visible before a total solar eclipse within

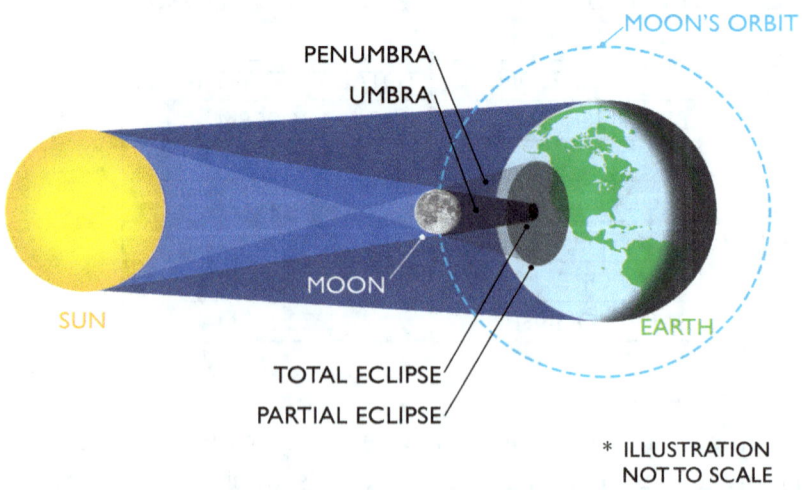

* ILLUSTRATION NOT TO SCALE

the inner shadow region. In both cases, it is imperative to use a solar filter on the lens for both photography and safety reasons. This is the only difference between taking a partial eclipse and a total eclipse photograph of the sun.

To photograph a total solar eclipse, you must be within the Path of Totality, the surface of the Earth within the moon's umbral shadow.

THE CHALLENGE

A total solar eclipse only lasts for a couple of minutes. It is brief, but the scenario it brings is unforgettable. Seeing the radiant sun slowly being covered by darkness gives the spectator a high level of anticipation and indescribable excitement. Once the moon completely covers the sun's radiance, the corona is finally visible. In the darkness, the sun's corona shines, capturing the crowd's full attention. Watching this phenomenon is a breathtaking experience.

Amidst all the noise, cheering, and excitement, you have no more than a few minutes to take a perfect photograph. The key to this is planning. You need to plan, practice, and perfect what you will do when the big moment arrives because there is no replay. The pressure is enormous. You only have a short time to capture the totality and the sun's corona using different exposures.

PLAN, PRACTICE, PERFECT

It is important to practice photographing before the actual phenomenon arrives. Test your chosen imaging setup for flaws. Rehearse over and over until your body remembers what you will do from the moment you arrive at your chosen spot to the moment you pack up and leave the area.

You will discover potential problems regarding vibrations and focus that you can address immediately. This minimizes the variables that might affect your photographs at the most critical moment.

It's common for experienced eclipse chasers to lose track of what they plan to do. Write down what you expect to do. Practice it time and again. Play annoying, distracting music while you practice. Try photographing in the worst weather possible. Do anything you can to practice under pressure. Eclipse day is not the time to practice.

Once the sun is completely covered, don't just take photographs. Capture the experience and the image of the total solar eclipse in your mind as well. Set up cameras around you to record not just the total solar eclipse but also the excitement and reaction of the crowd.

PHOTOGRAPHY

ECLIPSE PHOTOGRAPHY GEAR

What do you need to photograph the total eclipse? There are only a few pieces of equipment that you'll need. Preparing to photograph an eclipse successfully takes time. Not only do you have to be skilled and have the right gear, you have to be in the correct place.

BASIC ECLIPSE PHOTOGRAPHY EQUIPMENT

- Solar viewing glasses (verify authenticity)
- Lens solar filter
- Minimum 300mm lens
- Stable tripod that can be tilted to 60° vertical
- High-resolution DSLR
- Spare batteries for everything
- Secondary camera to photograph people, the horizon, etc.
- Remote cable or wireless release

ADDITIONAL ITEMS

- Video camera
- Video camera tripod
- Quality pair of binoculars
- Solar filters for each binocular lens
- Photo editing software

EQUIPMENT TO PREPARE BEFORE THE BIG DAY

A. Solar viewing glasses

You need a pair of solar viewing glasses as the eclipse approaches.

B. Solar Filter

Partial and total eclipse photography is different from normal photography. Even if only 1% of the sun's surface is visible, it is still approximately 10,000 times brighter than the moon. Before totality, use a solar filter on your lens. Do not look at the sun with your eyes. It can cause irreparable damage to your retinas.

DO NOT leave your camera pointed at the sun without a solar filter attached. The sun will melt the inside of your camera. Think of a magnifying glass used to torch ants and multiply that by one hundred.

C. Lens

To capture the corona's majesty, you need to use a telescope or a telephoto lens. The best focal length, which will give you a large image of the sun's disk, is 400mm and above. You don't want to waste all your efforts by bringing home a small dot where the black disk and majestic corona are supposed to be.

D. Tripod

Bring a stable enough tripod to support your camera properly to avoid unsteady shots and repeated adjustments. Either will ruin your photos. It also needs to be portable in case you need to change locations for a better shot. *Shut off camera stabilization on a tripod!*

E. Camera

You need to remember to set your camera to its highest resolution to capture all the details. Set your camera to:

- 14-bit RAW is ideal, otherwise
- JPG, Fine compression, Maximum resolution

Bracket your exposures. Shoot at various shutter speeds to capture different brightnesses in the corona. Note that stopping your lens all the way down may not result in the sharpest images.

Choose the lowest possible ISO for the best quality while maintaining a high shutter speed to prevent blurred shots. Set your camera to manual. Do not use AUTO ISO. Your camera will be fooled. The night before, test the focus position of your lens using a bright star or the moon.

Constantly double-check your focus. Be paranoid about this. You can deal with a grainy picture. No amount of Photoshop will fix a blurry, out-of-focus picture.

F. Batteries

Remember to bring fresh batteries! Make sure that you have enough power to capture the most important moments. Swap in fresh batteries thirty minutes before totality.

PHOTOGRAPHY

G. Remote release

Use a wired or wireless remote release to fire the camera's shutter. This will reduce the amount of camera vibration.

H. Video Camera

Run a video camera of yourself. Capture all the things you say and do during the totality. You'll be amazed at your reaction.

I. Photo editing software

You will need quality photo editing software to process your eclipse images. Adobe Lightroom and Photoshop are excellent programs to extract the most out of your images. Become well versed in how to use them at least a month before the eclipse.

J. Smartphone applications

The following smartphone applications will aid in your photography planning: Wunderground, Skyview, Photographer's Ephemeris, Sunrise and Sunset Calculator, SunCalc, and Sun Surveyor among others.

CAMERA PHONES

Smartphone cameras are useful for many things but not eclipse photography. An iPhone 6 camera has a 63° horizontal field of view and is 3264 pixels across. If you attempt to photograph the eclipse, the sun will be a measly 30-40 pixels wide depending on the phone. Digital pinch zoom won't help here. If you want *National Geographic* images, you'll need a serious camera and lens, far beyond any smartphone.

Consider instead using a smartphone to run a time-lapse of the entire event. The sun will be minuscule when shot on a smartphone. Think of something else exciting and interesting do to with it. Purchase a Gorilla Pod, inexpensive tripod, or selfie stick and mount the smartphone somewhere unique.

Also, partial and total eclipse light is strange and ethereal. Consider using that light to take unique pictures of things and people. It's rare and you may have something no one else does.

PHOTOGRAPHY

FOCAL LENGTH & THE SIZE OF SUN

The size of the sun in a photo depends on the lens focal length. A 300mm lens is the recommended minimum on a full-frame (FF) DSLR. Lenses up to this size are relatively inexpensive. For more magnification, use an APS-C (crop) size sensor. Cameras with these sensors provide an advantage by capturing a larger sun.

For the same focal length, an APS-C sensor will provide a greater apparent magnification of any object. As a consequence, a shorter, less expensive lens can be used to capture the same size sun.

The below figure shows the size of the sun on a camera sensor at various focal lengths. As can be seen with the 200mm lens, the sun is quite small. On a full-frame camera at 200mm, the sun will be 371 pixels wide on a Nikon D810, a 36-megapixel body. A lower resolution FF camera will result in an even smaller sun.

Printing a 24-inch image shot on a Nikon D810 with a 200mm lens at a standard 300 pixels per inch results in a small sun. On this size paper, the sun will be a miserly 1.25 inches wide!

Photographing the eclipse with a lens shorter than 300mm will leave you with little to work with. Using a 400mm lens and printing a 24-inch print will result in a 2.5-inch-wide sun. For as massive as the sun is, it is a challenge to take a large photograph of the sun. The sun will appear to move fast with a 500mm lens, too. Plan to adjust.

PHOTOGRAPHY

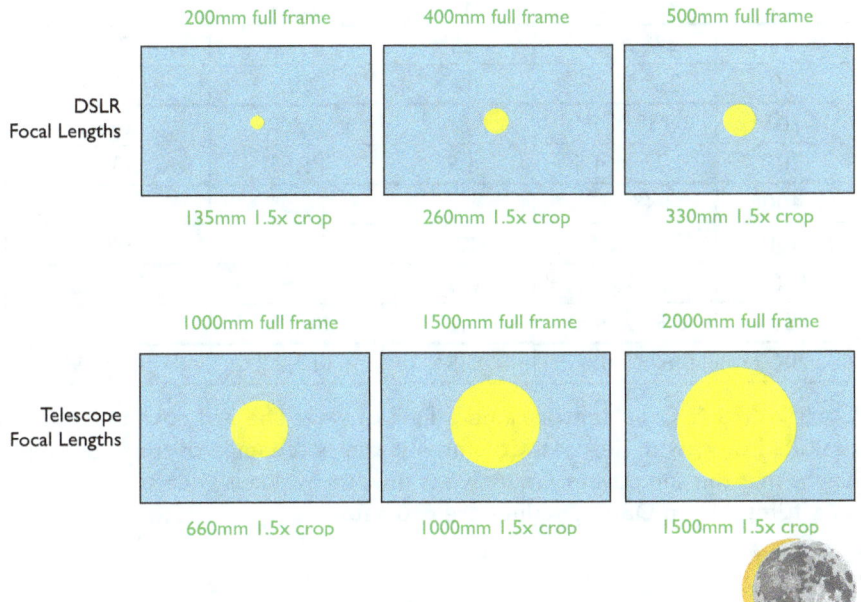

FOCAL LENGTH	FOV FULL FRAME	FF VERT. ANGLE	% OF FF	SUN PIXEL SIZE
14	104° X 81°	81°	0.7%	32.1
20	84° X 62°	62°	0.9%	41.9
28	65° X 46°	46°	1.2%	56.5
35	54° X 38°	38°	1.4%	68.5
50	40° X 27°	27°	2.0%	96.4
105	19° X 13°	13°	4.1%	200.2
200	10° X 7°	7°	7.6%	371.9
400	5° X 3.4°	3.4°	15.6%	765.6
500	4° X 2.7°	2.7°	19.6%	964.2
1000	2° X 1.3°	1.3°	40.8%	2002.5
1500	1.4° X 0.9°	0.9°	58.9%	2892.6
2000	1° X 0.68°	0.68°	77.9%	3828.4

Chart 1: Full-frame camera field of view. The 3rd column is the vertical field of view in degrees. Column 4 is the percentage of the total sensor height that the sun covers. Column 5 is how many pixels wide the sun will be on a 36MP Nikon D810. (Values are estimates)

FOCAL LENGTH	FOV CROP	CROP VERT DEG	% OF CROP	SUN PIXEL SIZE
14	80° X 58°	58°	0.9%	33.9
20	61° X 43°	43°	1.2%	45.8
28	45° X 31°	31°	1.7%	63.5
35	37° X 25°	25°	2.1%	78.7
50	26° X 18°	18°	2.9%	109.3
105	13° X 8°	8°	6.6%	245.9
200	6.7° X 4.5°	4.5°	11.8%	437.2
400	3.4° X 2°	2°	26.5%	983.7
500	2.7° X 1.8	1.8°	29.4%	1093.0
1000	1.3° X 0.9°	0.9°	58.9%	2186.0
1500	0.9° X 0.6°	0.6°	88.3%	3278.9
2000	0.6° X 0.45°	0.5°	117.8%	4371.9

Chart 2: APS-C Crop sensor camera field of view. The 3rd column is the vertical field of view in degrees. Column 4 is the percentage of the total sensor height that the sun covers. Column 5 is how many pixels wide the sun will be on a 12mp Nikon D300s. (Values are estimates)

PHOTOGRAPHY

The big challenge is the cost of the lens. Lenses longer than 300mm are expensive. They also require heavier tripods and specialized tripod heads. The 70-300mm lenses from Nikon, Canon, Tamron, and others are relatively affordable options. It is worth spending time at a local camera shop to try different lenses. Long focal-length lenses are a significant investment, especially for a single event.

To achieve a large eclipse image, you will need a long focal-length lens, ideally at least 400mm. A standard 70-300mm lens set to 300mm will show a small sun. At 500mm, the sun image becomes larger and covers more of the sensor area. The corona will take up a significant portion of the frame. By 1000mm, the corona will exceed the capture area on a full-frame sensor. See the picture in this chapter for sun size simulations for different focal lengths.

Suggested Exposures

To photograph the partial eclipse, the camera must have a solar filter attached. If not, the intense light from the sun may damage (fry) the inside of your camera. This has happened to the author. The exposure depends on the density (darkness) of the solar filter used.

As a starting point, set the camera to ISO 100, f/8, and with the solar filter on, try an exposure of 1/2000. Make adjustments based on the filter used, histogram, and highlight warning.

Turn on the highlight warning in your camera. This feature is commonly called "blinkies." This warning will help you detect if the image is overexposed or not.

Once the Baily's Beads, prominences, and corona become visible, there will only be a few minutes to take bracketed shots. It will take at least eleven shots to capture the various areas of the sun's corona and stars. The brightness varies considerably. No commercially available camera can capture the incredible dynamic range of the different portions of the delicate corona. This requires taking multiple photographs and digitally combining them afterward.

During totality, try these exposure times at ISO 100 and f/8: 1/4000, 1/2000, 1/1000, 1/250, 1/60, 1/30, 1/15, 1/4, 1/2, 1 sec, and 4 sec.

Disable camera/lens stabilization on a tripod!

PHOTOGRAPHY

PHOTOGRAPHY TIME

Set the camera to full-stop adjustments. It will reduce the time spent fiddling. As an example, the author tried the above shot sequence, adjusting the shutter speed as fast as possible.

It took thirty-three seconds to shoot the above 11 shots using 1/3-stop increments. This was without adjusting composition, focus, or anything else but the shutter speed. When the camera was set to full stop increments, it only took twenty-two seconds to step through the same shutter speed sequence. Use a remote release to reduce camera shake.

Assuming the totality lasts less than two minutes, only four shot sequences could be made using 1/3-stop increments. Yet six shot sequences could be made when the camera was set to full stop steps. Zero time was spent looking at the back LCD to analyze highlights and the histogram.

Now add in the bare minimum time to check the highlight warning. It took sixty-three seconds to shoot and check each image using full stops. And that was without changing the composition to allow for sun movement, bumping the tripod, etc. Looking at the LCD ("chimping") consumed **half** of the totality time.

This test was done in the comfort of home under no pressure. In real world conditions, it may be possible to successfully shoot only one sequence. If you plan to capture the entire dynamic range of the totality, you must practice the sequence until you have it down cold. If you normally fumble with your camera, do not underestimate the difficulty, frustration, and stress of total eclipse photography.

Most importantly, trying to shoot this sequence allowed for zero time to simply look at the totality to enjoy the spectacle.

AVOID LAST MINUTE PURCHASES

You should purchase whatever you think you'll need to photograph the eclipse early. This event will be nothing short of massive. Remember the hot toy of the year? Multiply that frenzy by a thousand. Everyone will want to try to capture their own photo.

Do not wait until the last few weeks before the eclipse to purchase cameras, lenses, filters, tripods, viewing glasses, and associated material. Consider that the totality of the eclipse will streak across

America. Everyone who wants to photograph the eclipse will order at the same time. If you wait until too late to buy what you need, it's conceivable that solar filters to create a total eclipse photo will be sold out in the United States. All filters sold out during the 2017 total eclipse. Whether this happens or not, do not wait to make your purchases. It may be too late.

PRACTICE

You will need to practice with your equipment. Things may go wrong that you don't anticipate. If you've never photographed a partial or total eclipse, taking quality shots is more difficult than you think. Practice shooting the sequence with a midday sun. This will tell you if you have your exposures and timing correct. Figure out what you need well in advance.

Practice photographing the full moon and stars at night. Capture the moon in full daylight to learn how your camera reacts. Astrophotography is challenging and requires practice.

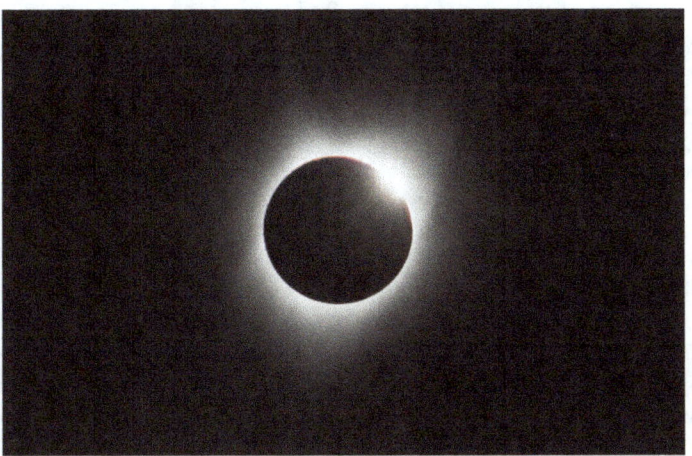

The August 21, 2017, eclipse as seen in Jackson, WY, shot with a Nikon D800 with an 80-400mm lens set to 340mm. The sun is 644 pixels wide on the 7360x4912 image.

This image is shown straight out of the camera without modification. Even with a high-quality camera and lens, photographing an eclipse is challenging.

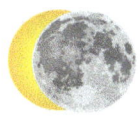

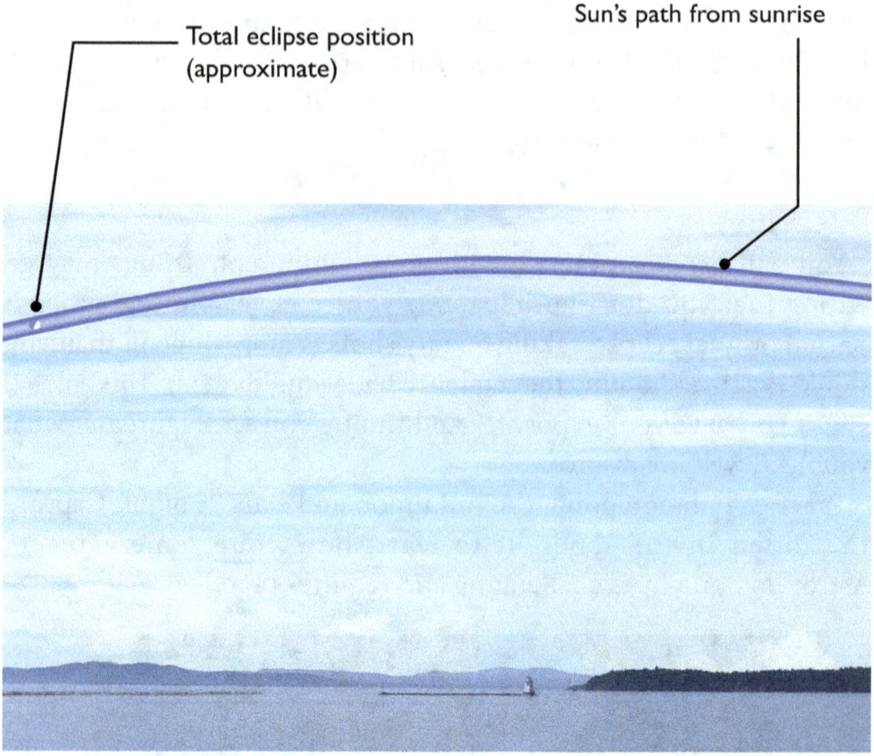

Total eclipse position
(approximate)

Sun's path from sunrise

The sun will follow this path on the morning of the eclipse on April 8, 2024. Image of Lake Champlain.

Note that this image is a simulation and approximation the sun's path and where the total eclipse may appear from one perspective. Refer to the eclipse position data for a more accurate location.

☉ is the symbol for the sun and first appeared in Europe during the Renaissance.
☾ is the ancient symbol for the moon.

LOCATIONS

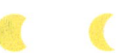

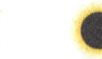

Viewing Locations Around Vermont

Tens of thousands of people will travel to and around Vermont to view the total eclipse. There are few obstructions and there is a vast amount of space to view the total eclipse from.

If the weather is questionable, seek out a new location as soon as possible. If you wait until the hour before the eclipse, you may find yourself stuck in traffic, as others will be looking for a viewing location. Be safe on the roadways, as drivers may be distracted.

This section contains popular, alternative, and little-known locations to watch the eclipse. As long as there are no clouds or smoke from fires, the partial eclipse will be viewable from anywhere in the state.

SUGGESTED TOTAL ECLIPSE VIEW POINTS

TOWNS AND CITIES

- Bakersfield
- Bristol
- Burlington
- Canaan
- Eden
- Enosburg
- Lyndon
- Middlebury
- Montpelier
- Newport
- Richford
- St. Albans City
- Saint Johnsbury
- Stowe
- Waitsfield

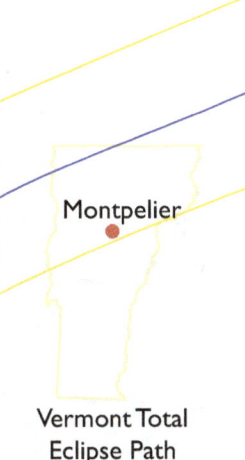

Vermont Total Eclipse Path

UNIQUE LOCATIONS

- Brighton State Park
- Jay State Forest
- Lake Carmi State Park
- Lake Champlain
- Lake Willoughby
- Mount Mansfield

LOCATIONS

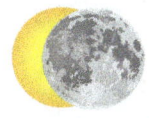

BAKERSFIELD

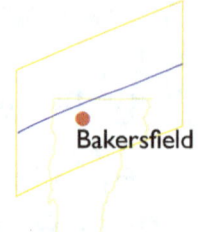

Bakersfield

Elevation:	751 ft. (229m)
Population:	1,322
Main road/hwy:	VT 36

OVERVIEW

Bakersfield is in the Islands and Farms region of the state and a short distance from Cold Hollow Sculpture Park, Lamoille Valley Rail Trail in Jeffersonville, and Saint Albans Museum. Stay overnight just thirteen minutes north at the 1906 House or twenty-two minutes west at the Back Inn Time Bed-and-Breakfast. Various dining options are a few minutes north and south of the city.

GETTING THERE

Drive north from Burlington on I-89, then continue east on VT 36 to reach Bakersfield.

TOTALITY DURATION

3 minutes 28 seconds

NOTES

Visit Bakersfield website for updated total eclipse event and lodging information at https://www.vermont.com/cities/bakersfield/.

Event	Time (EDT)	Altitude	Azimuth
Sunrise	6:18:00AM	0°	78°
Eclipse Start	2:14:58PM	48°	211°
Totality Start	3:26:30PM	40°	233°
Totality End	3:29:59PM	39°	234°
Eclipse End	4:37:35PM	29°	250°
Sunset	7:28:00PM	0°	282°

Bristol

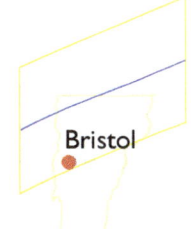

Elevation:	574 ft. (175m)
Population:	1,759
Main road/hwy:	US 19

Overview

Bristol is approximately an hour west of Montpelier and offers ski resorts for winter fun and golfing for summer leisure. Spend the afternoon in the cozy downtown area, have second helpings of artisan ice cream, and shop for a local souvenir to bring home. If you're lucky, you can catch the recycling picked up by horse and wagon. Don't miss Bristol Memorial Park Falls for a look at the waterfall or a hike on the trails. Adjacent to the park is the Bristol Cliffs Wilderness. Breathtaking rocky cliffs surround habitats for deer, bear, grouse, and falcons.

Getting There

Drive east from Burlington on I-89 for six miles, then turn south on VT 2A and continue for twenty-two miles to reach Bristol.

Totality Duration

1 minute 50 seconds

Notes

Learn more about the historic town of Bristol at https://www.vermontvacation.com/towns-and-regions/historic-downtowns/bristol.

Event	Time (EDT)	Altitude	Azimuth
Sunrise	6:20:00AM	0°	78°
Eclipse Start	2:14:10PM	49°	211°
Totality Start	3:26:52PM	40°	233°
Totality End	3:28:42PM	40°	234°
Eclipse End	4:37:26PM	29°	250°
Sunset	7:28:00PM	0°	281°

LOCATIONS

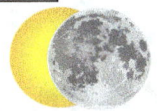

BURLINGTON

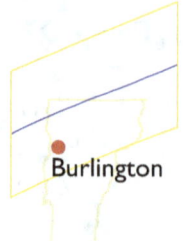

Burlington

Elevation:	190 ft. (58m)
Population:	42,556
Main road/hwy:	I-89

OVERVIEW

Burlington, the most populated city in the state, hugs Lake Champlain and offers a wide range of excellent hotels and dining options. Stroll Church Street Marketplace, an outdoor mall bustling with shops, outdoor entertainers, and intriguing architecture. Sign up for a walking tour, check out the aquarium, hop on the ferry around the lake, or rent a bike and enjoy a scenic trail with the kids. There are plenty of overnight accommodations here, so book a day or two before or after the eclipse and explore this charming city.

GETTING THERE

Burlington is the largest city in Vermont and can be reached by air (airport code: BTV), car, train from Washington D.C., and bus.

TOTALITY DURATION

3 minutes 15 seconds

NOTES

Start your exploration of Burlington at this website: https://vermont-vacation.com/towns-and-regions/historic-downtowns/burlington.

Event	Time (EDT)	Altitude	Azimuth
Sunrise	6:20:00AM	0°	78°
Eclipse Start	2:14:12PM	49°	211°
Totality Start	3:26:05PM	40°	233°
Totality End	3:29:20PM	40°	234°
Eclipse End	4:37:18PM	29°	250°
Sunset	7:29:00PM	0°	282°

CANAAN

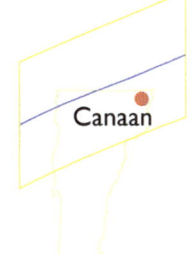

Elevation:	1,040 ft. (317m)
Population:	303
Main road/hwy:	US 3

OVERVIEW

Canaan in Essex County is a small, quaint village next to the Canadian border. Springtime in Vermont means the aroma of pure maple syrup will tempt your sweet tooth. Tour April's Maple house and enjoy a sample or three of fresh syrups, then bring home some treats from the gift shop. Lodging is limited, but try your luck at Jackson's Lodge or Maurice's Motel. Enjoy the eclipse with the friendly locals in Canaan.

GETTING THERE

Drive north from Montpelier on US 2 and VT 114N for ninety-six miles to reach Canaan.

TOTALITY DURATION

3 minutes 14 second

NOTES

Canaan's website is the perfect starting point for your total eclipse venture into northern Vermont at http://canaanvt.com/our-town/.

Event	Time (EDT)	Altitude	Azimuth
Sunrise	6:13:00AM	0°	78°
Eclipse Start	2:16:48PM	48°	214°
Totality Start	3:28:05PM	39°	235°
Totality End	3:31:19PM	38°	236°
Eclipse End	4:38:34PM	27°	251°
Sunset	7:23:00PM	0°	282°

LOCATIONS

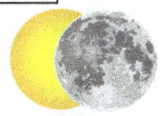

Eden

Elevation:	1,128 ft. (344m)
Population:	1,323
Main road/hwy:	VT 100 / VT 118

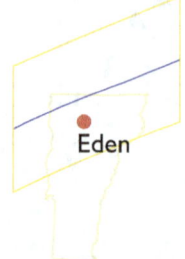

Overview

This little town is big on beauty. Once you've booked a room at the Fitchhill Inn less than twenty minutes away, Phinneas Swan Bed-and-Breakfast twenty minutes north, or Smuggler's Notch Inn just a few more minutes down the road, plan on hiking the Babcock Nature Preserve or picnic at the Lake Eden Recreation Area. Dining options nearby include Fledermaus Teahouse and Moog's Joint. The scenery in Eden is beautiful, so you'll want to take photos of the area before and after the eclipse.

Getting There

Drive north from Montpelier on VT 12 and continue on VT 100 for a total of thirty-nine miles to reach Eden.

Totality Duration

3 minutes 16 seconds

Notes

Visit the Eden town website at https://www.edenvt.org/ to start your journey into northern Vermont and learn more about the area.

LOCATIONS

Event	Time (EDT)	Altitude	Azimuth
Sunrise	6:17:00AM	0°	78°
Eclipse Start	2:15:16PM	48°	212°
Totality Start	3:26:53PM	40°	234°
Totality End	3:30:10PM	39°	234°
Eclipse End	4:37:49PM	28°	250°
Sunset	7:27:00PM	0°	281°

ENOSBURG

Elevation:	820 ft. (250m)
Population:	1,310
Main road/hwy:	VT 105

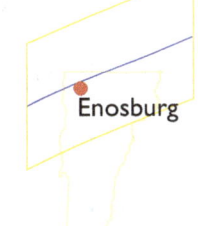

Enosburg

OVERVIEW

Enosburg is in northern Vermont, just below the Canadian border, along the Missisquoi River. Accommodations are limited, so book your room well in advance. While you're waiting for the big event, pick up lunch or dinner at one of the eateries along Main Street. Experience the Cold Hollow Sculpture Park, then venture out a half hour to Jay Peak Resort for skiing, golfing, hiking, or splashing around in the indoor waterpark. Head twenty minutes east and treat yourself to an afternoon at the Vermont Salt Cave Spa & Halotherapy Center.

GETTING THERE

Drive north from Burlington on I-89, then continue on VT 105 for a total of forty-six miles to reach Enosburg.

TOTALITY DURATION

3 minutes 32 seconds

NOTES

Enosburg's town website is a good jumping off place to learn more about this northern Vermont hamelet at http://enosburghvermont.org/.

Event	Time (EDT)	Altitude	Azimuth
Sunrise	6:18:00AM	0°	78°
Eclipse Start	2:15:03PM	48°	211°
Totality Start	3:26:30PM	40°	233°
Totality End	3:30:02PM	39°	234°
Eclipse End	4:37:34PM	29°	250°
Sunset	7:28:00PM	0°	281°

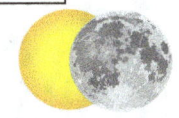

LYNDON

Elevation:	728 ft. (222m)
Population:	1,255
Main road/hwy:	I-91

OVERVIEW

The northeast corner of Vermont is refered to as the Northeast Kingdom. Nicknamed this over a half-century ago for the sheer natural splendor of the region, the name stuck and for good reason. This town in the Northeast Kingdom is just south of Willoughby State Forest. Book a room at Wildflower Inn, then head south about fifteen minutes to the Fairbanks Museum and Planetarium or the St. Johnsbury History and Heritage Center.

GETTING THERE

Drive northeast on US 2 from Montpelier, then continue north on I-91 for forty-four miles to reach Lyndon.

TOTALITY DURATION

2 minutes 17 seconds

NOTES

Explore Lyndon and the other communities in the Northeast Kingdom at https://www.vermont.com/northeast-kingdom/.

LOCATIONS

Event	Time (EDT)	Altitude	Azimuth
Sunrise	6:15:00AM	0°	78°
Eclipse Start	2:15:55PM	48°	213°
Totality Start	3:27:58PM	39°	235°
Totality End	3:30:16PM	39°	235°
Eclipse End	4:38:17PM	28°	251°
Sunset	7:24:00PM	0°	282°

MIDDLEBURY

Elevation:	416 ft. (127m)
Population:	6,922
Main road/hwy:	US 7

Middlebury

OVERVIEW

Middlebury sits at the southern tip of Lake Champlain and is listed on the National Register of Historic Places. Plan for a couple of days of pure charm. Take your camera when you stroll downtown Main Street. Otter Creek winds its way through the city and entices you along tidy sidewalks that will take you to restaurants, galleries, shops, and museums. Find a designated driver and hit the Middlebury Wine Tasting Trail, starting at Lincoln Peak Vineyard and ending at Windfall Orchard. You'll be sure to get a good night's sleep at one of the Bed-and-Breakfasts, and since Middlebury is utterly delightful, book early before everyone discovers this gem.

GETTING THERE

Drive south from Burlington on US 7 for thirty-four miles to reach Middlebury.

TOTALITY DURATION

58 seconds

NOTES

Visit Middlebury's website at https://experiencemiddlebury.com/.

LOCATIONS

Event	Time (EDT)	Altitude	Azimuth
Sunrise	6:20:00AM	0°	78°
Eclipse Start	2:13:58PM	49°	211°
Totality Start	3:27:10PM	40°	234°
Totality End	3:28:08PM	40°	234°
Eclipse End	4:37:23PM	29°	250°
Sunset	7:28:00PM	0°	282°

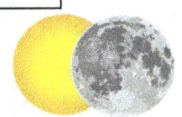

MONTPELIER

Elevation:	521 ft. (159m)
Population:	7,662
Main road/hwy:	I-89

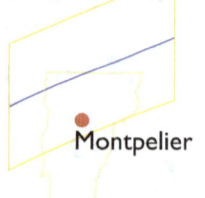

Montpelier

OVERVIEW

The capital city of Vermont is surrounded by things to do both indoors and outdoors. Start your visit at Hubbard Park just north of downtown, then follow the Winooski River as it winds through the city, leading you to museums, parks, an array of dining options, locally owned shops, and dropping you off at the golden-domed capitol building for a tour of the magnificent State House and art collection. Book ahead at the many lodging options such as the Inn at Montpelier and Doyles Guest House. Rest up as you'll need a couple of days to experience Montpelier's charm.

GETTING THERE

Montpelier is Vermont's capital city and can be reached by air (airport code MPV), car, or train.

TOTALITY DURATION

1 minute 36 seconds

NOTES

Learn more about Montpelier's golden-domed state house at https://statehouse.vermont.gov/.

Event	Time (EDT)	Altitude	Azimuth
Sunrise	6:17:00AM	0°	79°
Eclipse Start	2:14:57PM	49°	212°
Totality Start	3:27:36PM	40°	234°
Totality End	3:29:12PM	39°	235°
Eclipse End	4:37:51PM	29°	250°
Sunset	7:26:00PM	0°	281°

Newport

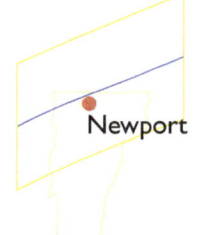

Elevation:	718 ft. (219m)
Population:	4,426
Main road/hwy:	US 5

Overview

Newport is just south the Canadian border, along US Route 5, near the shores of Lake Memphremagog. There's a lot to take in along the waterfront. Walk along Pomerleau Park, one of the stops in the historical walking tour, then head over to Jasper's Tavern or Eden's Specialty Ciders for a chilled beverage. Check out Newport City Dock and boardwalk, and lunch at the Newport Natural Chowder Shack. You'll have so much to do here that it will be night before you know it and time to rest up for another day.

Getting There

Drive north from Montpelier on VT 14, continue on VT 16, and then on to I-91, finally exiting on VT 191 to reach Newport for a total of sixty-three miles.

Totality Duration

3 minutes 25 seconds

Notes

Read more about Newport's recreational activities at https://www. newportrecreation.org/.

LOCATIONS

Event	Time (EDT)	Altitude	Azimuth
Sunrise	6:15:00AM	0°	78°
Eclipse Start	2:15:52PM	48°	212°
Totality Start	3:27:14PM	39°	234°
Totality End	3:30:40PM	39°	235°
Eclipse End	4:38:03PM	28°	250°
Sunset	7:26:00PM	0°	282°

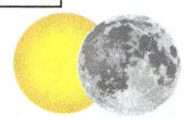

RICHFORD

Elevation:	465 ft. (142m)
Population:	1,397
Main road/hwy:	VT 105

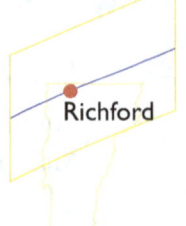

OVERVIEW

This small town is at the northern end of Vermont, just south of the Canadian border. Take a hike along the Missisquoi Valley Rail Trail or spend the day on the golf course. A few minutes south and to the west are forests worth exploring: the Berkshire Town Forest, Enosburg Village Forest, and to the west is Lake Carmi State Park where you can hike, swim, and camp right at the lake. Travel a half-hour drive along Route 105 to reach Jay State Forest for hiking and skiing. There is limited lodging in Richford. The Black Lantern Inn, just fifteen minutes south, is a good option. Dining at the Jay Village Inn and Restaurant, the Clubhouse Grille, or Alice's Table are worth a try.

GETTING THERE

Drive north from Burlington on I-89, exit and continue on VT 104/105 for a total of fifty-six miles to reach Richford.

TOTALITY DURATION

3 minutes 31 seconds

NOTES

Begin your exploration of the heart of the Green Mountains at Richford's website at http://richfordvt.org/.

Event	Time (EDT)	Altitude	Azimuth
Sunrise	6:17:00AM	0°	78°
Eclipse Start	2:15:17PM	48°	212°
Totality Start	3:26:40PM	39°	233°
Totality End	3:30:12PM	39°	234°
Eclipse End	4:37:40PM	28°	250°
Sunset	7:28:00PM	0°	282°

Saint Albans City

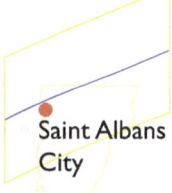

Elevation:	419 ft. (128m)
Population:	6,858
Main road/hwy:	US 7

Overview

Saint Albans City is a few minutes west of Lake Champlain along I-89 at the northern section of the state. Call ahead to book a room at the Back Inn Time Bed-and-Breakfast, a restored Victorian manor, and indulge in their wine and cheese package. Fill up on their home-cooked breakfast, then head west on Route 36 to Kamp Kill Kare State Park for swimming and boating. Enjoy a ferry ride to Burton Island State Park or just walk along Hathaway Point and take in the view. Come back to the downtown area for dinner along the walkable streets where you'll find plenty of restaurants, cafés, a historical museum, and interesting local shops.

Getting There

Drive north from Burlington on I-89 for twenty-nine miles and exit west on St. Albans State Highway to reach Saint Albans City.

Totality Duration

3 minutes 32 seconds

Notes

https://www.vermontvacation.com/towns-and-regions/historic-downtowns/st-albans

Event	Time (EDT)	Altitude	Azimuth
Sunrise	6:19:00AM	0°	78°
Eclipse Start	2:14:36PM	48°	211°
Totality Start	3:26:09PM	40°	233°
Totality End	3:29:42PM	39°	234°
Eclipse End	4:37:21PM	29°	250°
Sunset	7:29:00PM	0°	282°

LOCATIONS

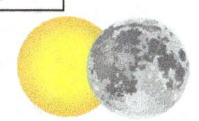

Saint Johnsbury

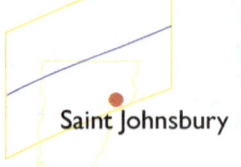

Elevation:	606 ft. (185m)
Population:	6,153
Main road/hwy:	I-93/I-91

Overview

Saint Johnsbury is an hour and a half east of Lake Champlain and has plenty to do while you enjoy your eclipse adventure. There are numerous hotel options so you can spend your entire visit right in town. Shop for local gifts, browse the galleries, or take in the varied architecture along the streets. Explore St. Johnsbury Athenaeum, Maple Grove Farms of Vermont, and the Fairbanks Museum and Planetarium that will feature eclipse-related events. Enjoy a variety of cuisines at one of the numerous restaurants, cafés, pubs, and taprooms. Go for a bike ride along the Lamoille Valley Rail Trail, then come back into town for a nightcap and a good night's sleep.

Getting There

Drive thirty-six miles northeast on US 2 from Montpelier to reach Saint Johnsbury.

Totality Duration

1 minute 31 seconds

Notes

Explore your travel options in this city at https://www.discoverst-johnsbury.com/.

Event	Time (EDT)	Altitude	Azimuth
Sunrise	6:15:00AM	0°	78°
Eclipse Start	2:15:49PM	48°	213°
Totality Start	3:28:19PM	39°	235°
Totality End	3:29:50PM	39°	235°
Eclipse End	4:38:17PM	28°	251°
Sunset	7:24:00PM	0°	282°

STOWE

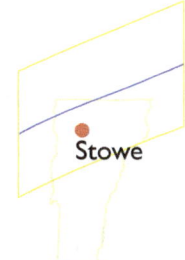

Elevation:	725 ft. (221m)
Population:	348
Main road/hwy:	VT 100

OVERVIEW

Options abound for accommodations in Stowe, a busy town less than an hour from the Burlington Airport. Book ahead at places like Stowe Meadows, Trapp Family Lodge, Green Mountain Inn, ski right into the Lodge at Spruce Peak, or rent a cabin in the woods. You'll find plenty of daytime fun in this mountain resort town. Take a snowmobile ride at ground level, zip tour above the trees, or go even higher in a hot air balloon. There are dining options for everyone's taste, from fancy to casual. After a day of adventure, treat yourself to a relaxing spa at Stoweflake Mountain Resort or get picked up from your hotel for a fun brewery tour from 4 Points VT. End your visit walking along Main Street, picking up crafts and gifts.

GETTING THERE

Drive northwest from Montpelier on I-89, then continue on VT 100 for a total of twenty-two miles to reach Stowe.

TOTALITY DURATION

2 minutes 48 seconds

NOTES

Visit Stowe's website at https://gostowe.com/.

Event	Time (EDT)	Altitude	Azimuth
Sunrise	6:18:00AM	0°	78°
Eclipse Start	2:14:55PM	49°	212°
Totality Start	3:26:54PM	40°	234°
Totality End	3:29:43PM	40°	234°
Eclipse End	4:37:44PM	29°	250°
Sunset	7:27:00PM	0°	281°

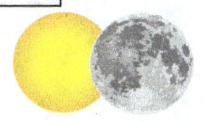

WAITSFIELD

Elevation:	695 ft. (212m)
Population:	168
Main road/hwy:	VT 100B

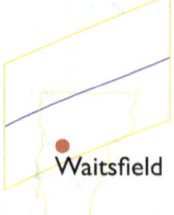

OVERVIEW

Waitsfield is a little town in the Mad River Valley with a host of activities for each season. During the eclipse, spring in the Mad River Valley still has mountain skiing, and mountain biking fever takes hold. Feeling adventurous? Fly glide with Sugarbush Soaring or keep closer to earth and sign up for a snowshoe maple sugaring tour. Walk along the Mad River Path, then visit the area's covered bridges. You'll want to take plenty of pictures to remember your stay in this scenic region of Vermont.

GETTING THERE

Drive west from Montpelier on I-89, then turn south on VT 100B for a total of nineteen miles to reach Waitsfield.

TOTALITY DURATION

1 minute 34 seconds

NOTES

Visit the Mad River Valley website at https://www.madrivervalley. com/ to learn more about visiting Waitsfield.

Event	Time (EDT)	Altitude	Azimuth
Sunrise	6:19:00PM	0°	78°
Eclipse Start	2:14:37PM	49°	212°
Totality Start	3:27:12PM	40°	234°
Totality End	3:28:56PM	40°	234°
Eclipse End	4:37:41PM	29°	250°
Sunset	7:27:00PM	0°	282°

Brighton State Park

Elevation: 1,187 ft. (362m)
Main road/hwy: VT 114

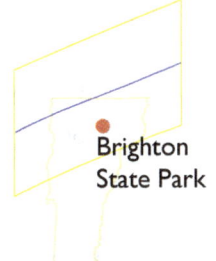

Brighton State Park

Overview

Brighton State Park, overlooking the gorgeous Island Pond, is a place fit for camping, fishing, and hiking. On the opposite side of the Pond, grab a bite to eat at Essex House and Tavern Cucina di Gerardo, or Jesse's Little Kitchen. Overnight accommodations are available approximately thirty minutes west of the park.

Getting There

Drive northwest on US 2 from Montpelier, continue north on I-91, then continue north on VT 114 for a total of sixty-nine miles to reach Brighton State Park.

Totality Duration

3 minutes 2 seconds

Notes

Visit Brighton State Park's webpage at https://vtstateparks.com/brighton.html for updated total eclipse event information.

LOCATIONS

Event	Time (EDT)	Altitude	Azimuth
Sunrise	6:14:00AM	0°	78°
Eclipse Start	2:16:16PM	48°	213°
Totality Start	3:27:48PM	49°	235°
Totality End	3:30:50PM	38°	235°
Eclipse End	4:38:21PM	28°	251°
Sunset	7:24:00PM	0°	282°

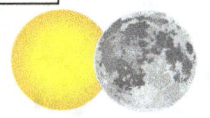

JAY STATE FOREST

Elevation:	3,763 ft. (1,147m)
Main road/hwy:	VT 242

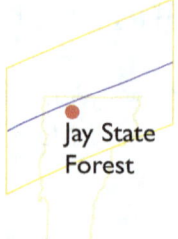

Jay State Forest

OVERVIEW

This forest is not too far from Westfield and split into two tracts: Black Falls and Big Jay. Head over to the Jay Peak ski resort or hike among the wildlife. The forest is a bird watcher's paradise as it is the natural habitat for several bird species, especially during spring and summer seasons. It is a small parcel filled with nature's gems.

GETTING THERE

Drive north from Montpelier for a total of sixty-two miles on VT 12, VT 100, VT 118, and VT 242 to reach Jay State Forest.

TOTALITY DURATION

3 minutes 30 seconds

NOTES

Start at the Jay State Forest webpage for more information on the area and any total eclipse events at https://fpr.vermont.gov/jay-state-forest-big-jayblack-falls-tracts.

Event	Time (EDT)	Altitude	Azimuth
Sunrise	6:17:00AM	0°	78°
Eclipse Start	2:15:26PM	48°	212°
Totality Start	3:26:50PM	39°	233°
Totality End	3:30:20PM	39°	234°
Eclipse End	4:37:48PM	28°	250°
Sunset	7:27:00PM	0°	281°

LOCATIONS

Lake Carmi State Park

Elevation: 442 ft. (135m)
Main road/hwy: VT 120

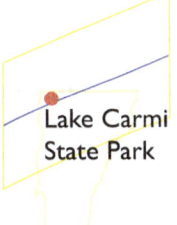

Lake Carmi State Park

Overview

Lake Carmi is the fourth-largest natural lake in Vermont at 1,375 acres. The lake is only thirty-three feet deep, so the lake supports warm-water species like walleyes and northern pike. The southern end of the lake is silted in, creating a unique bog area. The campground is substantial, with one-hundred thirty-eight tent and RV sites, with two cabins, and thirty-five lean-to-sites. The beach is also available for day use with restrooms and a nature center. The park interpreter is available for helping you learn more about the area.

Getting There

Drive north from Burlington on I-89, then continue on VT 105, then turn north on VT 236 for a total of forty-six miles to reach the state park.

Totality Duration

3 minutes 31 seconds

Notes

The Lake Carmi State Park webpage will have updated event information at https://vtstateparks.com/carmi.html.

Event	Time (EDT)	Altitude	Azimuth
Sunrise	6:18:00AM	0°	78°
Eclipse Start	2:15:00PM	48°	211°
Totality Start	3:26:26PM	40°	233°
Totality End	3:29:58PM	39°	234°
Eclipse End	4:37:30PM	29°	250°
Sunset	7:28:00PM	0°	282°

LOCATIONS

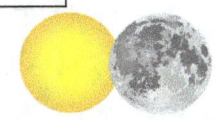

LAKE CHAMPLAIN

Elevation:	364 ft. (111m)
Main road/hwy:	Multiple

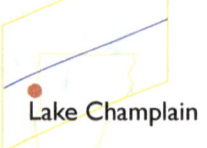

Lake Champlain

OVERVIEW

You'll need more than a day to discover all the surprises of Lake Champlain. Likened to a big aquarium or a maritime museum, this lake hosts several marine-life species unique to the region. Spring is the best time to visit the lake to enjoy outdoor activities like boating and hiking. Should April showers dampen your hike, there is plenty to do indoors. Take the kids to the Shelburne Museum and watch how Vermont Teddy Bears are made. Sample your heart out at Ben and Jerry's factory tour. If your feet need a break, take a ride on one of the ferry's three routes so you can sit back and take in the mountain view.

GETTING THERE

Lake Champlain is a substaintial lake on the northwest Vermont and New York border.

TOTALITY DURATION

Varies (*3 minutes 15 seconds for Burlington, VT*)

NOTES

https://www.vermont.org/experience/a-day-in-the-lake-champlain-islands

Times are for Burlington, VT

Event	Time (EDT)	Altitude	Azimuth
Sunrise	6:20:00AM	0°	78°
Eclipse Start	2:14:12PM	49°	211°
Totality Start	3:26:05PM	40°	233°
Totality End	3:29:20PM	40°	234°
Eclipse End	4:37:18PM	29°	250°
Sunset	7:29:00PM	0°	282°

Lake Willoughby

Elevation:	1,171 ft. (357m)
Main road/hwy:	VT 16

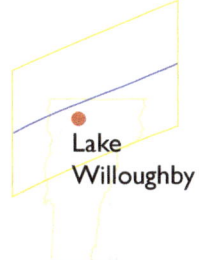

Overview

If you want to experience the eclipse away from crowds, keep this secret to yourself: drive over to Lake Willoughby in Westmore, Vermont, and soak up the scenery around the deepest lake in the state. Willoughby State Forest is on either side of the lake where you'll enjoy fishing, hiking, swimming, or canoeing. There are no camping areas inside the forest, so book a place in Westmore at the Willough Vale Inn, Will-O-Wood Campground, family-owned White Caps Campground, or Mountain Lake Cottages that offers a getaway package. Each lodging location offers dining options. Be cautious when you come to Lake Willoughby. It's quite breathtaking, and you may not want to leave.

Getting There

Drive east from Montpelier on US 2 to I-91, then turn north and continue to VT 16, then drive east to reach the lake.

Totality Duration

3 minutes 3 seconds

Notes

https://fpr.vermont.gov/willoughby-state-forest

Event	Time (EDT)	Altitude	Azimuth
Sunrise	6:15:00AM	0°	78°
Eclipse Start	2:15:58PM	48°	213°
Totality Start	3:27:33PM	39°	234°
Totality End	3:30:37PM	39°	235°
Eclipse End	4:38:12PM	28°	251°
Sunset	7:25:00PM	0°	281°

LOCATIONS

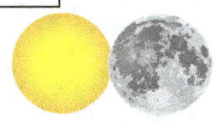

Mount Mansfield

Elevation:	4,258 ft. (1,298m)
Main road/hwy:	VT 108

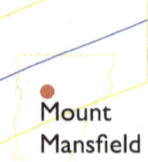

Mount
Mansfield

Overview

Mount Mansfield is the highest peak in the state. The surrounding forest encompasses Little River State Park, Smugglers' Notch State Park, Underhill State Park, and Waterbury Center State Park. Ski in the winter or hike in the summer and stay at one of the campgrounds for an up-close experience of rare old trees, amazing wildlife, and a wide diversity of flora and fauna. Bring your family to Little River State Park's popular campground for hiking, swimming, and boating. Trails on Mt. Mansfield are closed from mid-April to the end of May, but there are other trails to challenge your stamina.

Getting There

Drive east from Burlington on I-89, then continue north on VT 100 and VT 108 to Stowe Mountain Resort to reach Mount Mansfield.

Totality Duration

3 minutes 7 seconds

Notes

Visit Stowe's website as a starting point for hiking Mount Mansfield at https://gostowe.com/.

LOCATIONS

Event	Time (EDT)	Altitude	Azimuth
Sunrise	6:18:00AM	0°	78°
Eclipse Start	2:14:48PM	49°	211°
Totality Start	3:26:37PM	40°	233°
Totality End	3:29:45PM	39°	234°
Eclipse End	4:37:37PM	29°	250°
Sunset	7:27:00PM	0°	282°

REMEMBER THE VERMONT TOTAL ECLIPSE
April 8, 2024

Who was I with? _____
_____
_____
_____
_____

What did I see? _____
_____
_____
_____
_____
_____
_____
_____

What did I feel? _____
_____
_____
_____
_____
_____
_____

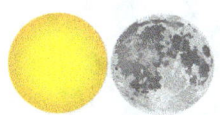

_____

_____

_____

_____

_____

What did the people with me think? _____

_____

_____

_____

_____

_____

_____

_____

_____

Where did I stay?_____

_____

_____

_____

_____

_____

_____

_____

_____

Enjoy other Sastrugi Press titles

2024 Total Eclipse State Series by Aaron Linsdau

Sastrugi Press has published state-specific guides for the 2024 total eclipse crossing over the United States. Check the Sastrugi Press website for the available state eclipse books: www.sastrugipress.com/eclipse

50 Wildlife Hotspots by Moose Henderson

Find out where to find animals and photograph them in Grand Teton National Park from a professional wildlife photographer. This unique guide shares the secret locations with the best chance at spotting wildlife.

A Small Pile of Feathers by Gerry Spence

Gerry Spence reveals his spiritual, loving, and sometimes humorous sides, depicted in his devotion to family and preserving the wild places he writes of as though they were inscribed on his own bones and in his own blood.

Adventure Expedition One by Aaron Linsdau, Terry Williams, M.D.

How do you plan and pull off your first epic expedition? Where should you even start? This book contains practical advice to help you put together your first trek. Dreaming, planning, training, doing, and returning alive are all covered in this guide.

Antarctic Tears by Aaron Linsdau

What would make someone give up a high-paying career to ski alone across Antarctica to the South Pole? This inspirational true story will make readers both cheer and cry. Fighting skin-freezing temperatures, infections, and emotional breakdown, Jackson Hole native Aaron Linsdau exposes the harsh realities of being on an expedition.

Cloudshade by Lori Howe, Ph.D.

The poems of *Cloudshade* breathe with the vivid, fragrant essence of life in every season on America's high plains. Extraordinarily relatable, the poems of *Cloudshade* swing wide a door to life in the West, both for lovers of poetry and for those who don't normally read poems.

Journeys to the Edge by Randall Peeters, Ph.D.

What is it like to climb Mount Everest? It requires dreaming big and creating a personal vision to climb the mountains in your life. Randall Peeters shares his successes and failures and provides the reader with some directly applicable guidelines on how to create a life vision.

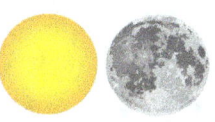

Lost at Windy Corner by Aaron Linsdau

Windy Corner on Denali has claimed lives, fingers, and toes. What would make some-one brave lethal weather, crevasses, and slick ice to attempt to summit North America's highest mountain? The author shares the lessons Denali teaches on managing goals and risks. Apply the message to build resilience and overcome adversity.

Sagebrush Alley by Patricia Jones

What's worse than having a stalker? Being pursued by a second one who has already killed. Attempting to complete her studies, Dana Cameron has to avoid becoming a murder victim. She becomes tangled in a struggle for life trapped in a claustrophobic nightmare.

Sleeping Dogs Don't Lie by Michael McCoy

A young Native American boy is taken from his home after tragedy strikes, grows up in middle America, and through his first real adult summer searches for Wyoming artifacts, falls in with the subversive Dog Soldiers Resurrected, and attempts single-handedly to solve the murder of his treasured coworker.

So I Said by Gerry Spence

The collected sayings of Gerry Spence provokes readers into thinking about their own vision of the world. As a lawyer with decades of experience in defending the defenseless, he's fought against giants. His insights provide a grander vision of how the nearly invisible world of the justice system in *So I Said*.

The Burqa Cave by Dean Petersen

Still haunted by Iraq, Tim Ross finds solace teaching high school in Wyoming. That is, until freshman David Jenkins reveals the murder of a lost local girl. Will Tim be able to overcome his demons to stop the murderer?

Voices at Twilight by Lori Howe, Ph.D.

Voices at Twilight is a guide takes readers on a visual tour of twelve past and present Wyoming ghost towns. Contained within are travel directions, GPS coordinates, and tips for intrepid readers.

Visit Sastrugi Press on the web at www.sastrugipress.com to purchase the above titles in bulk. They are also available from your local bookstore or online retailers in print, e-book, or audiobook form.

Thank you for choosing Sastrugi Press.

www.sastrugipress.com

"Turn the Page Loose"

About Aaron Linsdau, Polar Explorer & Motivational Speaker

Aaron Linsdau is a polar explorer and motivational speaker. He energizes audiences with life and business lessons that stick. He is an expert at building resilience to overcome adversity by maintaining a positive attitude. Aaron teaches audiences how to eat two sticks of butter a day to achieve their goal. He shares how to deal with constant pressure, burnout, and adrenaline overload.

He holds the world record for the longest expedition in days from Hercules Inlet to the South Pole. Aaron is the second only American to complete the trip alone, eating seventy pounds of butter on the expedition.

This solo expedition is more difficult than climbing Mount Everest with a team. Being alone dramatically increases the challenge. Aaron uses emotionally stirring stories to show how to overcome adversity, manage risk and safety, and survive unimaginable conditions. He relates these stories to business challenges and shows how the common person can achieve uncommon results.

Aaron collaborates with organizations to deliver the right message for the audience. He connects his stories to business realities. Aaron loves inspiring audiences. Book Aaron for your next event today.

"Never Give Up"
Adversity • Attitude • Resilience • Risk • Safety

Read reviews, be inspired, and learn more about Aaron Linsdau at:
www.aaronlinsdau.com

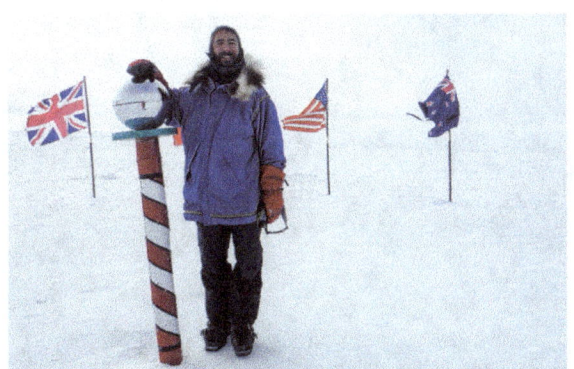

Aaron at the South Pole after 82 days alone in Antarctica.

Smartphone link

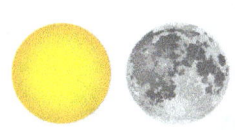